An Introduction to Nonparametric Statistics

CHAPMAN & HALL/CRC
Texts in Statistical Science Series
Joseph K. Blitzstein, *Harvard University, USA*
Julian J. Faraway, *University of Bath, UK*
Martin Tanner, *Northwestern University, USA*
Jim Zidek, *University of British Columbia, Canada*

Recently Published Titles

The Analysis of Time Series
An Introduction with R, Seventh Edition
Chris Chatfield and Haipeng Xing

Time Series
A Data Analysis Approach Using R
Robert H. Shumway and David S. Stoffer

Practical Multivariate Analysis, Sixth Edition
Abdelmonem Afifi, Susanne May, Robin A. Donatello, and Virginia A. Clark

Time Series: A First Course with Bootstrap Starter
Tucker S. McElroy and Dimitris N. Politis

Probability and Bayesian Modeling
Jim Albert and Jingchen Hu

Surrogates
Gaussian Process Modeling, Design, and Optimization for the Applied Sciences
Robert B. Gramacy

Statistical Analysis of Financial Data
With Examples in R
James Gentle

Statistical Rethinking
A Bayesian Course with Examples in R and STAN, Second Edition
Richard McElreath

Statistical Machine Learning
A Model-Based Approach
Richard Golden

Randomization, Bootstrap and Monte Carlo Methods in Biology
Fourth Edition
Bryan F. J. Manly, Jorje A. Navarro Alberto

Principles of Uncertainty
Second Edition
Joseph B. Kadane

An Introduction to Nonparametric Statistics
John E. Kolassa

For more information about this series, please visit: https://www.crcpress.com/Chapman--HallCRC-Texts-in-Statistical-Science/book-series/CHTEXSTASCI

An Introduction to Nonparametric Statistics

John E. Kolassa

CRC Press
Taylor & Francis Group
Boca Raton London New York

CRC Press is an imprint of the
Taylor & Francis Group, an **informa** business

A CHAPMAN & HALL BOOK

First edition published 2021
by CRC Press
6000 Broken Sound Parkway NW, Suite 300, Boca Raton, FL 33487-2742

and by CRC Press
2 Park Square, Milton Park, Abingdon, Oxon, OX14 4RN

© 2021 Taylor & Francis Group, LLC

CRC Press is an imprint of Taylor & Francis Group, LLC

Reasonable efforts have been made to publish reliable data and information, but the author and publisher cannot assume responsibility for the validity of all materials or the consequences of their use. The authors and publishers have attempted to trace the copyright holders of all material reproduced in this publication and apologize to copyright holders if permission to publish in this form has not been obtained. If any copyright material has not been acknowledged please write and let us know so we may rectify in any future reprint.

Except as permitted under U.S. Copyright Law, no part of this book may be reprinted, reproduced, transmitted, or utilized in any form by any electronic, mechanical, or other means, now known or hereafter invented, including photocopying, microfilming, and recording, or in any information storage or retrieval system, without written permission from the publishers.

For permission to photocopy or use material electronically from this work, access www.copyright.com or contact the Copyright Clearance Center, Inc. (CCC), 222 Rosewood Drive, Danvers, MA 01923, 978-750-8400. For works that are not available on CCC please contact mpkbookspermissions@tandf.co.uk

Trademark notice: Product or corporate names may be trademarks or registered trademarks, and are used only for identification and explanation without intent to infringe.

Library of Congress Cataloging-in-Publication Data
Names: Kolassa, John Edward, 1963- author.
Title: An introduction to nonparametric statistics / John Kolassa.
Description: First edition. | Boca Raton : CRC Press, 2020. |
Series: Chapman & Hall/CRC texts in statistical science |
Includes bibliographical references and index.
Identifiers: LCCN 2020023408 (print) | LCCN 2020023409 (ebook) | ISBN
   9780367194840 (hardback) | ISBN 9780429202759 (ebook)
Subjects: LCSH: Nonparametric statistics.
Classification: LCC QA278.8 .K65 2020 (print) | LCC QA278.8
   (ebook) | DDC 519.5--dc23
LC record available at https://lccn.loc.gov/2020023408
LC ebook record available at https://lccn.loc.gov/2020023409

ISBN: 9780367194840 (hbk)
ISBN: 9780429202759 (ebk)

Typeset in CMR
by Nova Techset Private Limited, Bengaluru & Chennai, India

Contents

Introduction			**xi**
1 Background			**1**
1.1	Probability Background		1
	1.1.1 Probability Distributions for Observations		1
		1.1.1.1 Gaussian Distribution	1
		1.1.1.2 Uniform Distribution	2
		1.1.1.3 Laplace Distribution	3
		1.1.1.4 Cauchy Distribution	3
		1.1.1.5 Logistic Distribution	4
		1.1.1.6 Exponential Distribution	4
	1.1.2 Location and Scale Families		4
	1.1.3 Sampling Distributions		5
		1.1.3.1 Binomial Distribution	5
	1.1.4 χ^2-distribution		5
	1.1.5 T-distribution		6
	1.1.6 F-distribution		6
1.2	Elementary Tasks in Frequentist Inference		6
	1.2.1 Hypothesis Testing		6
		1.2.1.1 One-Sided Hypothesis Tests	7
		1.2.1.2 Two-Sided Hypothesis Tests	9
		1.2.1.3 P-values	10
	1.2.2 Confidence Intervals		11
		1.2.2.1 P-value Inversion	11
		1.2.2.2 Test Inversion with Pivotal Statistics	12
		1.2.2.3 A Problematic Example	12
1.3	Exercises		13
2 One-Sample Nonparametric Inference			**15**
2.1	Parametric Inference on Means		15
	2.1.1 Estimation Using Averages		15
	2.1.2 One-Sample Testing for Gaussian Observations		16
2.2	The Need for Distribution-Free Tests		16
2.3	One-Sample Median Methods		17
	2.3.1 Estimates of the Population Median		18
	2.3.2 Hypothesis Tests Concerning the Population Median .		19

		2.3.3 Confidence Intervals for the Median	24

        2.3.3 Confidence Intervals for the Median 24
        2.3.4 Inference for Other Quantiles 27
  2.4 Comparing Tests 28
        2.4.1 Power, Sample Size, and Effect Size 29
                2.4.1.1 Power 29
                2.4.1.2 Sample and Effect Sizes 30
        2.4.2 Efficiency Calculations 31
        2.4.3 Examples of Power Calculations 33
  2.5 Distribution Function Estimation 34
  2.6 Exercises 35

3 Two-Sample Testing   39
  3.1 Two-Sample Approximately Gaussian Inference 39
        3.1.1 Two-Sample Approximately Gaussian Inference on Expectations 39
        3.1.2 Approximately Gaussian Dispersion Inference 40
  3.2 General Two-Sample Rank Tests 41
        3.2.1 Null Distributions of General Rank Statistics 41
        3.2.2 Moments of Rank Statistics 42
  3.3 A First Distribution-Free Test 43
  3.4 The Mann-Whitney-Wilcoxon Test 46
        3.4.1 Exact and Approximate Mann-Whitney Probabilities 48
                3.4.1.1 Moments and Approximate Normality 48
        3.4.2 Other Scoring Schemes 50
        3.4.3 Using Data as Scores: the Permutation Test 51
  3.5 Empirical Levels and Powers of Two-Sample Tests 53
  3.6 Adaptation to the Presence of Tied Observations 54
  3.7 Mann-Whitney-Wilcoxon Null Hypotheses 55
  3.8 Efficiency and Power of Two-Sample Tests 55
        3.8.1 Efficacy of the Gaussian-Theory Test 55
        3.8.2 Efficacy of the Mann-Whitney-Wilcoxon Test 56
        3.8.3 Summarizing Asymptotic Relative Efficiency 57
        3.8.4 Power for Mann-Whitney-Wilcoxon Testing 57
  3.9 Testing Equality of Dispersion 58
  3.10 Two-Sample Estimation and Confidence Intervals 60
        3.10.1 Inversion of the Mann-Whitney-Wilcoxon Test 61
  3.11 Tests for Broad Alternatives 62
  3.12 Exercises 64

4 Methods for Three or More Groups   69
  4.1 Gaussian-Theory Methods 69
        4.1.1 Contrasts 70
        4.1.2 Multiple Comparisons 71
  4.2 General Rank Tests 73
        4.2.1 Moments of General Rank Sums 73

	4.2.2	Construction of a Chi-Square-Distributed Statistic	74
4.3		The Kruskal-Wallis Test	76
	4.3.1	Kruskal-Wallis Approximate Critical Values	76
4.4		Other Scores for Multi-Sample Rank Based Tests	78
4.5		Multiple Comparisons	80
4.6		Ordered Alternatives	82
4.7		Powers of Tests	84
	4.7.1	Power of Tests for Ordered Alternatives	84
	4.7.2	Power of Tests for Unordered Alternatives	85
4.8		Efficiency Calculations	90
	4.8.1	Ordered Alternatives	91
	4.8.2	Unordered Alternatives	91
4.9		Exercises	92

5 Group Differences with Blocking   95

5.1		Gaussian Theory Approaches	95
	5.1.1	Paired Comparisons	95
	5.1.2	Multiple Group Comparisons	95
5.2		Nonparametric Paired Comparisons	96
	5.2.1	Estimating the Population Median Difference	98
	5.2.2	Confidence Intervals	100
	5.2.3	Signed-Rank Statistic Alternative Distribution	101
5.3		Two-Way Non-Parametric Analysis of Variance	102
	5.3.1	Distribution of Rank Sums	102
5.4		A Generalization of the Test of Friedman	103
	5.4.1	The Balanced Case	104
	5.4.2	The Unbalanced Case	105
5.5		Multiple Comparisons and Scoring	107
5.6		Tests for a Putative Ordering in Two-Way Layouts	108
5.7		Exercises	110

6 Bivariate Methods   113

6.1		Parametric Approach	113
6.2		Permutation Inference	114
6.3		Nonparametric Correlation	115
	6.3.1	Rank Correlation	115
		6.3.1.1 Alternative Expectation of the Spearman Correlation	118
	6.3.2	Kendall's τ	118
6.4		Bivariate Semi-Parametric Estimation via Correlation	121
	6.4.1	Inversion of the Test of Zero Correlation	121
		6.4.1.1 Inversion of the Pearson Correlation	121
		6.4.1.2 Inversion of Kendall's τ	122
		6.4.1.3 Inversion of the Spearman Correlation	123
6.5		Exercises	126

7 Multivariate Analysis — 129
- 7.1 Standard Parametric Approaches — 129
  - 7.1.1 Multivariate Estimation — 129
  - 7.1.2 One-Sample Testing — 129
  - 7.1.3 Two-Sample Testing — 130
- 7.2 Nonparametric Multivariate Estimation — 130
  - 7.2.1 Equivariance Properties — 132
- 7.3 Nonparametric One-Sample Testing Approaches — 132
  - 7.3.1 More General Permutation Solutions — 135
- 7.4 Confidence Regions for a Vector Shift Parameter — 136
- 7.5 Two-Sample Methods — 136
  - 7.5.1 Hypothesis Testing — 137
    - 7.5.1.1 Permutation Testing — 137
    - 7.5.1.2 Permutation Distribution Approximations — 140
- 7.6 Exercises — 141

8 Density Estimation — 143
- 8.1 Histograms — 143
- 8.2 Kernel Density Estimates — 144
- 8.3 Exercises — 148

9 Regression Function Estimates — 149
- 9.1 Standard Regression Inference — 149
- 9.2 Kernel and Local Regression Smoothing — 150
- 9.3 Isotonic Regression — 154
- 9.4 Splines — 155
- 9.5 Quantile Regression — 157
  - 9.5.1 Fitting the Quantile Regression Model — 159
- 9.6 Resistant Regression — 162
- 9.7 Exercises — 165

10 Resampling Techniques — 167
- 10.1 The Bootstrap Idea — 167
  - 10.1.1 The Bootstrap Sampling Scheme — 168
- 10.2 Univariate Bootstrap Techniques — 170
  - 10.2.1 The Normal Method — 170
  - 10.2.2 Basic Interval — 170
  - 10.2.3 The Percentile Method — 171
  - 10.2.4 BC_a Method — 172
  - 10.2.5 Summary So Far, and More Examples — 174
- 10.3 Bootstrapping Multivariate Data Sets — 175
  - 10.3.1 Regression Models and the Studentized Bootstrap Method — 176
  - 10.3.2 Fixed X Bootstrap — 178
- 10.4 The Jackknife — 181

	10.4.1 Examples of Biases of the Proper Order	181
	10.4.2 Bias Correction	182
	10.4.2.1 Correcting the Bias in Mean Estimators	182
	10.4.2.2 Correcting the Bias in Quantile Estimators	182
10.5	Exercises	185

A Analysis Using the SAS System   187

B Construction of Heuristic Tables and Figures Using R   197

Bibliography   201

Index   211

Introduction

Preface

This book is intended to accompany a one-semester MS-level course in nonparametric statistics. Prerequisites for the course are calculus through multivariate Taylor series, elementary matrix algebra including matrix inversion, and a first course in frequentist statistical methods including some basic probability. Most of the techniques described in this book apply to data with only minimal restrictions placed on their probability distributions, but performance of these techniques, and the performance of analogous parametric procedures, depend on these probability distributions. The first chapter below reviews probability distributions. It also reviews some objectives of standard frequentist analyses. Chapters covering methods that have elementary parametric counterparts begin by reviewing those counterparts. These introductions are intended to give a common terminology for later comparisons with new methods, and are not intended to reflect the richness of standard statistical analysis, or to substitute for an intentional study of these techniques.

Computational Tools and Data Sources

Conceptual developments in this text are intended to be independent of the computational tools used in practice, but analyses used to illustrate techniques developed in this book will be facilitated using the program R (R Core Team, 2018). This program may be downloaded for free from https://cran.r-project.org/ . This course will heavily depend on the R package MultNonParam. This package is part of the standard R repository CRAN, and is installed by typing inside of R:

library(MultNonParam)

Other packages will be called as needed; if your system does not have these installed, install them as above, substituting the package name for MultNonParam.

Calculations of a more heuristic nature, and not intended for routine analysis, are performed using the package NonparametricHeuristic. Since this package is so tightly tied to the presentation in this book, and hence of less general interest, it is hosted on the github repository, and installed via

```
library(devtools)
install_github("kolassa-dev/NonparametricHeuristic")
```

and, once installed, loaded into R using the library command.

An appendix gives guidance on performing some of these calculations using SAS.

Errata and other materials will be posted at

http://stat.rutgers.edu/home/kolassa/NonparametricBook

as they become available.

Acknowledgments

In addition to works referenced in the following chapters, I consulted Stigler (1986) and Hald (1998) for early bibliographical references. Bibliographic trails have been tracked through documentation of software packages R and SAS, and bibliography resources from JSTOR, Citulike.org, Project Euclid, and various publishers' web sites have been used to construct the bibliography. I am grateful to the Rutgers Physical Sciences librarian Melanie Miller, the Rutgers Department of Statistics Administrative Assistant Lisa Curtin, and the work study students that she supervised, and the Rutgers Interlibrary Loan staff for assistance in locating reference material.

I am grateful to my students at both Rutgers and the University of Rochester to whom I taught this material over the years. Halley Constantino, Jianning Yang, and Peng Zhang used a preliminary version of this manuscript and were generous with their suggestions for improvements. My experience teaching this material helped me to select material for this volume, and to determine its level and scope. I consulted various textbooks during this time, including those of Hettmansperger and McKean (2011), Hettmansperger (1984), and Higgins (2004).

I thank my family for their patience with me during preparation of this volume. I thank my editor and proofreader for their important contributions.

I dedicate this volume to my wife, Yodit.

1
Background

Statistics is the solution to an inverse problem: given the outcome from a random process, the statistician infers aspects of the underlying probabilistic structure that generated the data. This chapter reviews some elementary aspects of probability, and then reviews some classical tools for inference about a distribution's location parameter.

1.1 Probability Background

This section first reviews some important elementary probability distributions, and then reviews a tool for embedding a probability distribution into a larger family that allows for the distribution to be recentered and rescaled. Most statistical techniques described in this volume are best suited to continuous distributions, and so all of these examples of plausible data sources are continuous.

1.1.1 Probability Distributions for Observations

Some common probability distributions are shown in Figure 1.1. The continuous distributions described below might plausibly give rise to a data set of independent observations. This volume is intended to direct statistical inference on a data set without knowing the family from which it came. The behavior of various statistical procedures, including both standard parametric analyses, and nonparametric techniques forming the subject of this volume, may depend on the distribution generating the data, and knowledge of these families will be used to explore this behavior.

1.1.1.1 Gaussian Distribution

The normal distribution, or Gaussian distribution, has density

$$f_G(x) = \exp(-(x-\mu)^2/(2\sigma^2))/(\sigma\sqrt{2\pi}).$$

FIGURE 1.1: Comparison of Three Densities

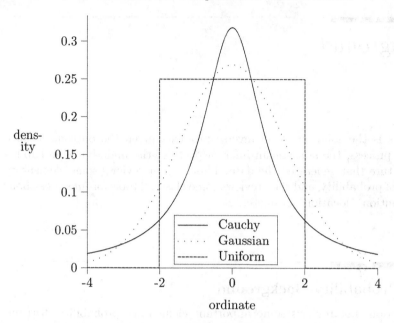

The parameter μ is both the expectation and the median, and σ is the standard deviation. The Gaussian cumulative distribution function is

$$F_G(x) = \int_{-\infty}^{x} f_G(y)\,dy.$$

There is no closed form for this integral. This distribution is symmetric about μ; that is, $f_G(x) = f_G(2\mu - x)$, and $F_G(x) = 1 - F_G(2\mu - x)$. The specific member of this family of distributions with $\mu = 0$ and $\sigma = 1$ is called standard normal distribution or standard Gaussian distribution. The standard Gaussian cumulative distribution function is denoted by $\Phi(x)$. The distribution in its generality will be denoted by $\mathfrak{G}(\mu, \sigma^2)$.

This Gaussian distribution may be extended to higher dimensions; a multivariate Gaussian random variable, or multivariate normal random variable, $\boldsymbol{X}$, in a space of dimension d, has a density of form $\exp(-(\boldsymbol{x} - \boldsymbol{\mu})^\top \boldsymbol{\Upsilon}^{-1}(\boldsymbol{x} - \boldsymbol{\mu})/2) \det \boldsymbol{\Upsilon}^{-1/2} (2\pi)^{-d/2}$. Here $\boldsymbol{\mu}$ is the expectation $\mathrm{E}\left[\boldsymbol{X}\right]$, and $\boldsymbol{\Upsilon}$ is the variance-covariance matrix $\mathrm{E}\left[(\boldsymbol{X} - \boldsymbol{\mu})^\top (\boldsymbol{X} - \boldsymbol{\mu})\right]$.

1.1.1.2 Uniform Distribution

The uniform distribution has density

$$f_U(x) = \begin{cases} 1/\lambda & \text{for } x \in (\theta - \lambda/2, \theta + \lambda/2) \\ 0 & \text{otherwise} \end{cases}.$$

Probability Background

The cumulative distribution function of this distribution is

$$F_U(x) = \begin{cases} 0 & \text{for } x \leq \theta - \lambda/2 \\ 1/2 + (x - \theta)/\lambda & \text{for } x \in (\theta - \lambda/2, \theta + \lambda/2) \\ 1 & \text{for } x \geq \theta + \lambda/2 \end{cases}.$$

Again, the expectation and median for this distribution are both θ, and the distribution is symmetric about θ. The standard deviation is $\lambda/\sqrt{12}$. A common canonical member of this family is the distribution uniform on $[0,1]$, with $\theta = 1/2$ and $\lambda = 1$. The distribution in its generality will be denoted by $\mathcal{U}(\theta, \lambda)$.

1.1.1.3 Laplace Distribution

The <u>double exponential distribution</u> or <u>Laplace distribution</u> has density

$$f_{La}(x) = \exp(-|x - \theta|\sqrt{2}/\sigma))/(\sigma\sqrt{2}).$$

The cumulative distribution function for this distribution is

$$F_{La}(x) = \begin{cases} \exp((x - \theta)\sqrt{2}/\sigma))/2 & \text{for } x \leq \theta \\ 1 - \exp(-(x - \theta)\sqrt{2}/\sigma))/2 & \text{f } x > \theta \end{cases}.$$

As before, the expectation and median of this distribution are both θ. The standard deviation of this distribution is σ. The distribution is symmetric about θ. A canonical member of this family is the one with $\theta = 0$ and $\sigma = 1$. The distribution in its generality will be denoted by $\mathcal{L}a(\theta, \sigma^2)$.

1.1.1.4 Cauchy Distribution

Consider the family of distributions

$$f_C(x) = \varsigma^{-1}/(\pi(1 + (x - \theta)^2/\varsigma^2)), \tag{1.1}$$

with θ real and ς positive. The cumulative distribution function for such a distribution is

$$F_C(x) = 1/2 + \arctan((x - \theta)/\varsigma)/\pi.$$

This distribution is symmetric about its median θ, but, unlike the Gaussian, uniform, and Laplace examples, does not have either an expectation nor a variance; the quantity ς represents not a standard deviation but a more general scaling parameter. Its upper and lower quartiles are $\theta \pm \varsigma$, and so the interquartile range is 2ς. This distribution is continuous. The family member with $\theta = 0$ and $\varsigma = 1$ is the <u>Cauchy distribution</u> (and, when necessary to distinguish it from other members (1.1), will be called standard Cauchy), and a member with $\varsigma = 1$ but $\theta \neq 0$ is called a <u>non-central Cauchy distribution</u>.

An interesting and important property of the Cauchy relates to the distribution of sums of independent and identical copies of members of this family. If X and Y are independent standard Cauchy, then $Z = (X + Y)/2$

is standard Cauchy. One may see this by first noting that $P[Z \leq z] = \int_{-\infty}^{\infty} \int_{-\infty}^{2z-y} f_C(x) f_C(y) \, dx \, dy$, and hence that $f_Z(z) = \int_{-\infty}^{\infty} f_C(2z-y) f_C(y) \, dy$. Dwass (1985) evaluates this integral using partial fractions. Alternatively, this fact might be verified using characteristic functions.

1.1.1.5 Logistic Distribution

The <u>logistic distribution</u> has density

$$f_L(x) = (1 + \exp(-(x-\theta)/\sigma))^{-1}(1 + \exp((x-\theta)/\sigma))^{-1} \sigma^{-1}.$$

This distribution is symmetric about θ, and has expectation and variance θ and $\sigma^2 \pi^2/3$ respectively. The cumulative distribution function of this distribution is

$$F_L(x) = \exp((x-\theta)/\sigma)/(1 + \exp((x-\theta))).$$

The distribution in its generality will be denoted by $\mathfrak{Lo}(\theta, \sigma^2)$.

1.1.1.6 Exponential Distribution

The <u>exponential distribution</u> has density

$$f_E(x) = \begin{cases} \exp(-x) & \text{for } x > 0 \\ 0 & \text{otherwise} \end{cases}.$$

The cumulative distribution function of this distribution is

$$F_E(x) = \begin{cases} 0 & \text{for } x \leq 0 \\ 1 - \exp(-x) & \text{for } x > 0 \end{cases}.$$

The expectation is 1, and the median is $\log(2)$. The inequality of these values is an indication of the asymmetry of the distribution. The standard deviation is 1. The distribution will be denoted by $\mathfrak{E}$.

1.1.2 Location and Scale Families

Most of the distributions above are or can be naturally extended to a family of distributions, called a <u>location-scale family</u>, by allowing an unknown location constant to shift the center of the distribution, and a second unknown scale constant to shrink or expand the scale. That is, suppose that X has density $f(x)$ and cumulative distribution function $F(x)$. Then $a + bX$ has density $f((y-a)/b)/b$ and the cumulative distribution function $F((y-a)/b)$. If X has a standard distribution with location and scale 0 and 1, then Y has location a and scale b.

This argument does not apply to the exponential distribution, because the lower endpoint of the support of the distribution often is fixed at zero by the structure of the application in which it is used.

Probability Background 5

1.1.3 Sampling Distributions

The distributions presented above in §1.1.1 represent mechanisms for generating observations that might potentially be analyzed nonparametricly. Distributions in this subsection will be used in this volume primarily as sampling distributions, or approximate sampling distributions, of test statistics.

1.1.3.1 Binomial Distribution

The binomial distribution will be of use, not to model data plausibly arising from it, but because it will be used to generate the first of the nonparametric tests considered below. This distribution is supported on $\{0, 1, \ldots, n\}$ for some integer n, and has an additional parameter $\pi \in [0, 1]$. Its probability mass function is $\binom{n}{x}\pi^x(1-\pi)^{n-x}$, and its cumulative distribution function is

$$F_B(x) = \sum_{y=0}^{x} \binom{n}{y}\pi^y(1-\pi)^{n-y}.$$

Curiously, the binomial cumulative distribution function can be expressed in terms of the cumulative distribution function of the F distribution, to be discussed below. The expectation is $n\pi$, and the variance is $n\pi(1-\pi)$. The median does not have a closed-form expression. This distribution is symmetric only if $\pi = 1/2$. The distribution will be denoted by $\mathfrak{Bin}(n, \pi)$.

The multinomial distribution extends the binomial distribution to the distribution of counts of objects independently classified into more than two categories according to certain probabilities.

1.1.4 χ^2-distribution

If $X_1, \ldots, X_k$ are independent random variables, each with a standard Gaussian distribution (that is, Gaussian with expectation zero and variance one), then the distribution of the sum of their squares is called the chi-square distribution, and is denoted χ_k^2. Here the index k is called the degrees of freedom.

Distributions of quadratic forms of correlated summands sometimes have a χ^2 distribution as well. If $\boldsymbol{Y}$ has a multivariate Gaussian distribution with dimension k, expectation $\boldsymbol{0}$ and variance matrix $\boldsymbol{\Upsilon}$, and if $\boldsymbol{\Upsilon}$ has an inverse, then

$$\boldsymbol{Y}^\top \boldsymbol{\Upsilon}^{-1} \boldsymbol{Y} \sim \chi_k^2.$$

One can see this by noting that $\boldsymbol{\Upsilon}$ may be written as $\boldsymbol{\Theta}\boldsymbol{\Theta}^\top$. Then $\boldsymbol{X} = \boldsymbol{\Theta}^{-1}\boldsymbol{Y}$ is multivariate Gaussian with expectation $\boldsymbol{0}$, and variance matrix $\boldsymbol{\Theta}^{-1}\boldsymbol{\Theta}\boldsymbol{\Theta}^\top\boldsymbol{\Theta}^{-1\top} = \boldsymbol{I}$, where $\boldsymbol{I}$ is the identity matrix with k rows and columns. Then $\boldsymbol{X}$ is a vector of independent standard Gaussian variables, and $\boldsymbol{Y}^\top \boldsymbol{\Upsilon}^{-1}\boldsymbol{Y} = \boldsymbol{X}^\top \boldsymbol{X}$.

Furthermore, still assuming

$$X_j \sim \mathfrak{G}(0, 1), \text{ independent} \tag{1.2}$$

the distribution of

$$W = \sum_{i=1}^{k}(X_i - \delta_i)^2 \qquad (1.3)$$

is called a non-central chi-square distribution (Johnson et al., 1995). The density and distribution function of this distribution are complicated. The most important property of this distribution is that it depends on $\delta_1, \ldots, \delta_k$ only through $\xi = \sum_{i=1}^{k} \delta_k^2$; this quantity is known as the non-centrality parameter, and the distribution of W will be denoted by $\chi_k^2(\xi)$. This dependence on nonzero expectations only through the simple non-centrality parameter may be seen by calculating the moment generating function of this distribution.

If $\boldsymbol{Y}$ has a multivariate Gaussian distribution with dimension k, expectation $\boldsymbol{0}$ and variance matrix $\boldsymbol{\Upsilon}$, and if $\boldsymbol{\Upsilon}$ has an inverse, then $\boldsymbol{X} = \boldsymbol{\Theta}^{-1}(\boldsymbol{Y} - \boldsymbol{\mu})$ is multivariate Gaussian with expectation $-\boldsymbol{\Theta}^{-1}\boldsymbol{\mu}$, and variance matrix $\boldsymbol{I}$. Hence

$$(\boldsymbol{Y} - \boldsymbol{\mu})^\top \boldsymbol{\Upsilon}^{-1}(\boldsymbol{Y} - \boldsymbol{\mu}) \sim \chi_k^2(\xi) \text{ with } \xi = \boldsymbol{\mu}^\top \boldsymbol{\Upsilon}^{-1}\boldsymbol{\mu}. \qquad (1.4)$$

1.1.5 T-distribution

When U has a standard Gaussian distribution, V has a χ_k^2 distribution, and U and V are independent, then $T = U/\sqrt{V/k}$ has a distribution called Student's t distribution, denoted here by $\mathfrak{T}_k$, with k called the degrees of freedom.

1.1.6 F-distribution

When U and V are independent random variables, with χ_k^2 and χ_m^2 distributions respectively, then $F = (U/k)/(V/m)$ is said to have an F distribution with k and m degrees of freedom; denote this distribution by $\mathfrak{F}_{k,m}$. If a variable T has a $\mathfrak{T}_m$ distribution, then T^2 has an $\mathfrak{F}_{1,m}$ distribution.

1.2 Elementary Tasks in Frequentist Inference

This section reviews elementary frequentist tasks of hypothesis testing and producing confidence intervals.

1.2.1 Hypothesis Testing

Statisticians are often asked to choose among potential hypotheses about the mechanism generating a set of data. This choice is often phrased as between a null hypothesis, generally implying the absence of a potential effect, and an alternative hypothesis, generally the presence of a potential effect. These

Elementary Tasks in Frequentist Inference

hypotheses in turn are expressed in terms of sets of probability distributions, or equivalently, in terms of restrictions on a summary (such as the median) of a probability distribution. A hypothesis test is a rule that takes a data set and returns "Reject the null hypothesis that $\theta = \theta^0$" or "Do not reject the null hypothesis."

In many applications studied in this manuscript, hypothesis tests are implemented by constructing a test statistic T, depending on data, and a constant t°, such that the statistician rejects the null hypothesis (1.6) if $T \geq t^\circ$, and fails to reject it otherwise. The constant t° is called a critical value, and the collection of data sets in

$$\{\text{data} | T(\text{data}) \geq t^\circ\} \tag{1.5}$$

for which the null hypothesis is rejected is called the critical region.

1.2.1.1 One-Sided Hypothesis Tests

For example, if data represent the changes in some physiological measure after receiving some therapy, measured on subjects acting independently, then a null hypothesis might be that each of the changes in measurements comes from a distribution with median zero, and the alternative hypothesis might be that each of the changes in measurements comes from a distribution with median greater than zero. In symbols, if θ represents the median of the distribution of changes, then the null hypothesis is $\theta = 0$, and the alternative is

$$\theta \geq 0. \tag{1.6}$$

Such an alternative hypothesis is typically called a one-sided hypothesis. The null hypothesis might then be thought of as the set of all possible distributions for observations with median zero, and the alternative is the set of all possible distributions for observations with positive median. If larger values of θ make larger values of T more likely, and smaller values less likely, then a test rejecting the null hypothesis for data in (1.5) is reasonable.

Because null hypotheses are generally smaller in dimension than alternative hypotheses, frequencies of errors are generally easier to control for null hypotheses than for alternative hypotheses.

Tests are constructed so that, in cases in which the null hypothesis is actually true, it is rejected with no more than a fixed probability. This probability is called the test level or type I error rate, and is commonly denoted α.

Hence the critical value in such cases is defined to be the smallest value t° satisfying

$$P_{\theta^0}[T \geq t^\circ] \leq \alpha. \tag{1.7}$$

When the distribution of T is continuous then the critical value satisfies (1.7) with $\leq$ replaced by equality. Many applications in this volume feature test statistics with a discrete distribution; in this case, the $\leq$ in (1.7) is generally $<$.

The other type of possible error occurs when the alternative hypothesis is true, but the null hypothesis is not rejected. The probability of erroneously failing to reject the null hypothesis is called the type II error rate, and is denoted β. More commonly, the behavior of the test under an alternative hypothesis is described in terms of the probability of a correct answer, rather than of an error; this probability is called power; power is $1 - \beta$.

One might attempt to control power as well, but, unfortunately, in the common case in which an alternative hypothesis contains probability distributions arbitrarily close to those in the null hypothesis, the type II error rate will come arbitrarily close to one minus the test level, which is quite large. Furthermore, for a fixed sample size and mechanism for generating data, once a particular distribution in the alternative hypothesis is selected, the smallest possible type II error rate is fixed, and cannot be independently controlled. Hence generally, tests are constructed primarily to control level. Under this paradigm, then, a test is constructed by specifying α, choosing a critical value to give the test this type I error, and determining whether this test rejects the null hypothesis or fails to reject the null hypothesis.

In the one-sided alternative hypothesis formulation $\{\theta > \theta^0\}$, the investigator is, at least in principle, interested in detecting departures from the null hypothesis that vary in proximity to the null hypothesis. (The same observation will hold for two-sided tests in the next subsection). For planning purposes, however, investigators often pick a particular value within the alternative hypothesis. This particular value might be the minimal value of practical interest, or a value that other investigators have estimated. They then calculate the power at this alternative, to ensure that it is large enough to meet their needs. A power that is too small indicates that there is a substantial chance that the investigator's alternative hypothesis is correct, but that they will fail to demonstrate it. Powers near 80% are typical targets.

Consider a test with a null hypothesis of form $\theta = \theta^0$ and an alternative hypothesis of form $\theta = \theta_A$, using a statistic T such that under the null hypothesis $T \sim \Phi(\theta^0, \sigma_0^2)$, and under the alternative hypothesis $T \sim \Phi(\theta_A, \sigma_A^2)$. Test with a level α, and without loss of generality assume that $\theta_A > \theta^0$. In this case, the critical value is approximately $t^\circ = \theta^0 + \sigma_0 z_\alpha$. Here z_α is the number such that $\Phi(z_\alpha) = 1 - \alpha$. A common level for such a one-sided test is $\alpha = 0.025$; $z_{0.025} = 1.96$. The power for such a one-sided test is

$$1 - \Phi((t^\circ - \theta_A)/\sigma_A) = 1 - \Phi((\theta^0 + \sigma_0 z_\alpha - \theta_A)/\sigma_A). \qquad (1.8)$$

One might plan an experiment by substituting null hypothesis values θ^0 and σ_0 and alternative hypothesis values θ_A and σ_A into (1.8), and verifying that this power is high enough to meet the investigator's needs; alternatively, one might require power to be $1 - \beta$, and solve for the effect size necessary to give this power. This effect size is

$$\theta_A = \theta^0 + \sigma_A z_\beta + \sigma_0 z_\alpha. \qquad (1.9)$$

Elementary Tasks in Frequentist Inference 9

One can then ask whether this effect size is plausible. More commonly, σ_0 and σ_A both may be made to depend on a parameter representing sample size, with both decreasing to zero as sample size increases. Then (1.8) is solved for sample size.

1.2.1.2 Two-Sided Hypothesis Tests

In contrast to one-sided alternatives, consider a two-sided hypothesis for $\theta \ne \theta^0$, and, in this case, one often uses a test that rejects the null hypothesis for data sets in

$$\{T \le t_L^\circ\} \cup \{T \ge t_U^\circ\}. \tag{1.10}$$

In this case, there are two critical values, chosen so that

$$\alpha \ge \mathrm{P}_{\theta^0}\left[T \le t_L^\circ \text{ or } T \ge t_U^\circ\right] = \mathrm{P}_{\theta^0}\left[T \le t_L^\circ\right] + \mathrm{P}_{\theta^0}\left[T \ge t_U^\circ\right], \tag{1.11}$$

and so that the first inequality is close to equality.

Many of the statistics constructed in this volume require an arbitrary choice by the analyst of direction of effect, and choosing the direction of effect differently typically changes the sign of T. In order keep the analytical results invariant to this choice of direction, the critical values are chosen to make the two final probabilities in (1.11) equal. Then the critical values solve the equations

$$\mathrm{P}_{\theta^0}\left[T \ge t_U^\circ\right] \le \alpha/2, \text{ and } \mathrm{P}_{\theta^0}\left[T \le t_L^\circ\right] \le \alpha/2, \tag{1.12}$$

with t_U° chosen as small as possible, and t_L° chosen as large as possible consistent with (1.12). Comparing with (1.7), the two-sided critical value is calculated exactly as is the one-sided critical value for an alternative hypothesis in the appropriate direction, and for a test level half that of the one-sided test. Hence a two-sided test of level 0.05 is constructed in the same way as two one-sided tests of level 0.025.

Often the critical region implicit in (1.11) can be represented by creating a new statistic that is large when T is either large or small. That is, one might set $W = |T - \mathrm{E}_{\theta^0}[T]|$. In this case, if t_L° and t_U° are symmetric about $\mathrm{E}_{\theta^0}[T]$, then the two-sided critical region might be expressed as

$$\{W \ge w^\circ\} \tag{1.13}$$

for $w^\circ = t_U^\circ - \mathrm{E}_{\theta^0}[T]$. Alternatively, one might define $W = |T - \mathrm{E}_{\theta^0}[T]|^2$, and, under the same symmetry condition, use as the critical region (1.13) for $w^\circ = (t_U^\circ - \mathrm{E}_{\theta^0}[T])^2$. In the absence of such a symmetry condition, w° may be calculated from the distribution of W directly, by choosing w° to make the probability of the set in (1.13) equal to α.

Statistics T for which (1.10) is a reasonable critical region are inherently one-sided, since the two-sided test is constructed from one-sided tests combining evidence pointing in opposite directions. Similarly, statistics W for which (1.13) is a reasonable critical region for the two-sided alternative are inherently two-sided.

Power for the two-sided test is the same probability as calculated in (1.11), with the θ_A substituted for θ^0, and power substituted for α. Again, assume that large values of θ make larger values of T more likely. Then, for alternatives θ_A greater than θ^0, the first probability added in

$$1 - \beta = P_{\theta_A}[T \leq t_L^\circ] + P_{\theta_A}[T \geq t_U^\circ]$$

is quite small, and is typically ignored for power calculations. Additionally, rejection of the null hypothesis because the evidence is in the opposite direction of that anticipated will result in conclusions from the experiment not comparable to those for which the power calculation is constructed. Hence power for the two-sided test is generally approximated as the power for the one-sided test with level half that of the two-sided tests, and α in (1.8) is often 0.025, corresponding to half of the two-sided test level.

Some tests to be constructed in this volume may be expressed as $W = \sum_{j=1}^{k} U_j^2$ for variables U_j which are, under the null hypothesis, approximately standard Gaussian and independent; furthermore, in such cases, the critical region for such tests is often of the form $\{W \geq w^\circ\}$. In such cases, the test of level α rejects the null hypothesis when $W \geq \chi_{k,\alpha}^2$, for $\chi_{k,\alpha}^2$ the $1 - \alpha$ quantile of the χ_k^2 distribution. If the standard Gaussian approximation for the distribution of U_j is only approximately correct, then the resulting test will have level α approximately, but not exactly.

If, under the alternative hypothesis, the variables U_j have expectations μ_j and standard deviations ε_j, the alternative distribution of W will be a complicated weighted sum of $\chi_1^2(\mu_j^2)$ variables. Usually, however, the impact of the move from the null distribution to the alternative distribution is much higher on the component expectations than on the standard deviations, and one might treat these alternative standard deviations fixed at 1. With this simplification, the sampling distribution of W under the alternative is $\chi_k^2(\sum_{i=1}^{k} \delta_i^2)$, the non-central chi-square distribution.

1.2.1.3 P-values

Alternatively, one might calculate a test statistic, and determine the test level at which one transitions from rejecting to not rejecting the null hypothesis. This quantity is called a *p-value*. For a one-sided test with critical region of form (1.5), the p-value is given by

$$P_0[T \geq t_{\text{obs}}], \tag{1.14}$$

for t_{obs} the observed value of the test statistic. For two-sided critical values of form (1.10), with condition (1.12), the p-value is given by

$$2\min(P_0[T \geq t_{\text{obs}}], P_0[T \leq t_{\text{obs}}]). \tag{1.15}$$

These p-values are interpreted as leading to rejection of the null hypothesis when they are as low as or lower than the test level specified in advance by the investigator before data collection.

Elementary Tasks in Frequentist Inference 11

Inferential procedures that highlight *p*-values are indicative of the inferential approach of Fisher (1925), while those that highlight pre-specified test levels and powers are indicative of the approach of Neyman and Pearson (1933).

I refer readers to a thorough survey (Lehmann, 1993), and note here only that while I find the pre-specified test level arguments compelling, problematic examples leading to undesirable interpretations of *p*-values are rare using the techniques developed in this volume, and, more generally, the contrasts between techniques advocated by these schools of thought are not central to the questions investigated here.

1.2.2 Confidence Intervals

A confidence interval of level $1 - \alpha$ for parameter θ is defined as a set (L, U) such that L and U depend on data, and such that for any θ,

$$P_\theta [L < \theta < U] \geq 1 - \alpha.$$

The most general method for constructing a confidence interval is test inversion. For every possible null value θ^0, find a test of the null hypothesis $\theta = \theta^0$, versus the two-sided alternative, of level no larger than α. Then the confidence set is

$$\{\theta^0 | \text{The null hypothesis } \theta = \theta^0 \text{ is not rejected.}\}. \quad (1.16)$$

In many cases, (1.16) is an interval. In such cases, one attempts to determine the lower and upper bounds of the interval, either analytically or numerically.

Often, such tests are phrased in terms of a quantity $W(\theta)$ depending on both the data and the parameter, such that the test rejects the null hypothesis $\theta = \theta^0$ if and only if $W(\theta^0) \geq w^\circ(\theta^0)$ for some critical value c that might depend on the null hypothesis.

1.2.2.1 *P*-value Inversion

One might construct confidence intervals through tail probability inversion. Suppose that one can find a univariate statistic T whose distribution depends on the unknown parameter θ, such that potential one-sided *p*-values are monotonic in θ for each potential statistic value t. Typical applications have

$$P_\theta [T \geq t] \text{ nondecreasing in } \theta, \ P_\theta [T \leq t] \text{ non-increasing in } \theta \ \forall t, \quad (1.17)$$

with probabilities in (1.17) continuous in θ. Let t be the observed value of T. Under (1.17),

$$\{\theta | P_\theta [T \geq t] > \alpha/2, P_\theta [T \leq t] > \alpha/2\}$$

is an interval, of form (θ^L, θ^U), with endpoints satisfying

$$P_{\theta^L} [T \geq t] = \alpha/2, P_{\theta^U} [T \leq t] = \alpha/2. \quad (1.18)$$

There may be t such that the equation $P_{\theta^L}[T \geq t] = \alpha/2$ has no solution, because $P_\theta[T \geq t] > \alpha/2$ for all θ. In such cases, take θ^L to be the lower bound on possible values for θ. For example, if $\pi \in [0,1]$, and $T \sim \mathrm{Bin}(n, \pi)$, then $P_\pi[T \geq 0] = 1 > \alpha/2$ for all π, $P_\pi[T \geq 0] = \alpha/2$ has no solution, and $\pi^L = 0$. Alternatively, if θ can take any real value, and $T \sim \mathrm{Bin}(n, \exp(\theta)/(1+\exp(\theta)))$, then $P_\theta[T \geq 0] = \alpha/2$ has no solution, and $\theta^L = -\infty$. Similarly, there may be t such that the equation $P_{\theta^U}[T \leq t] = \alpha/2$ has no solution, because $P_\theta[T \leq t] > \alpha/2$ for all θ. In such cases, take θ^U to be the upper bound on possible values for θ.

Construction of intervals for the binomial proportion represents a simple example in which p-values may be inverted (Clopper and Pearson, 1934).

1.2.2.2 Test Inversion with Pivotal Statistics

Confidence interval construction is simplified when there exists a random quantity, generally involving an unknown parameter θ, with a distribution that does not depend on θ. Such a quantity is called a pivot. For instance, in the case of independent and identically distributed observations with average $\bar{X}$ and standard deviation s from a $\mathfrak{G}(\theta, \sigma^2)$ distribution, then $T = (\bar{X} - \theta)/(s/\sqrt{n})$ has a t distribution with $n - 1$ degrees of freedom, regardless of θ.

One may construct a confidence interval using a pivot by finding quantiles t_L° and t_U° such that

$$P[t_L^\circ < T(\theta, \text{data}) < t_U^\circ] \geq 1 - \alpha. \tag{1.19}$$

Then

$$\{\theta | t_L^\circ < T(\theta, \text{data}) < t_U^\circ\} \tag{1.20}$$

is a confidence interval, if it is really an interval. In the case when (1.20) is an interval, and when $T(\theta, \text{data})$ is continuous in θ, then the interval is of the form (L, U); that is, the interval does not include the endpoints.

1.2.2.3 A Problematic Example

One should use this test inversion technique with care, as the following problematic case shows. Suppose that X and Y are Gaussian random variables, with expectations μ and ν respectively, and common known variances σ^2. Suppose that one desires a confidence interval for $\rho = \mu/\nu$ (Fieller, 1954). The quantity $T = \sqrt{n}(\bar{X} - \rho\bar{Y})/(\sigma\sqrt{1+\rho^2})$ has a standard Gaussian distribution, independent of ρ, and hence is pivotal. A confidence region is $\{\rho : n(\bar{X} - \rho\bar{Y})^2/\sigma^2(1+\rho^2)) \leq z_{\alpha/2}^2\}$. Equivalently, the region is

$$\{\rho | Q(\rho) < 0\} \text{ for } Q(\rho) = (X^2 - \upsilon^2)\rho^2 - 2XY\rho + Y^2 - \upsilon^2 \tag{1.21}$$

for $\upsilon = \sigma z_{\alpha/2}$.

If $X^2 + Y^2 < \upsilon^2$, then $Q(\rho)$ in (1.21) has a negative coefficient for ρ^2, and the maximum value is at $\rho = XY/(X^2 - \upsilon^2)$. The maximum is

… *Exercises*

$(v^2(-v^2 + X^2 + Y^2))/(v^2 - X^2) < 0$, and so the inequality in (1.21) holds for all ρ, and the confidence interval is the entire real line.

If $X^2 + Y^2 > v^2 > X^2$, then the quadratic form in (1.21) has a negative coefficient for ρ^2, and the maximum is positive. Hence values satisfying the inequality in (1.21) are very large and very small values of ρ; that is, the confidence interval is

$$\left(-\infty, \frac{-XY - v\sqrt{X^2 + Y^2 - v^2}}{v^2 - X^2}\right) \cup \left(\frac{-XY + v\sqrt{X^2 + Y^2 - v^2}}{v^2 - X^2}, \infty\right).$$

If $X^2 > v^2$, then the quadratic form in (1.21) has a positive coefficient for ρ^2, and the minimum is negative. Then the values of ρ satisfying the inequality in (1.21) are those near the minimizer $XY/(X^2 - v^2)$. Hence the interval is

$$\left(\frac{XY - v\sqrt{X^2 + Y^2 - v^2}}{X^2 - v^2}, \frac{XY + v\sqrt{X^2 + Y^2 - v^2}}{X^2 - v^2}\right).$$

1.3 Exercises

1. Demonstrate that the moment generating function for the statistic (1.3), under (1.2), depends on $\delta_1, \ldots, \delta_k$ only through $\sum_{j=1}^{k} \delta_j^2$.

2

One-Sample Nonparametric Inference

This chapter first reviews standard Gaussian-theory inference on one sample location models. It then presents motivation for why a distribution-free approach to location testing is necessary, and presents nonparametric techniques for inference on quantiles. Later in this chapter, techniques for comparing the efficiencies of tests are introduced, and these are applied to various parametric and nonparametric tests. Finally, techniques for estimating a single cumulative distribution function are discussed.

2.1 Parametric Inference on Means

Suppose one wants to learn about $\theta = \mathrm{E}[X_j]$, from a sample $X_1, \ldots, X_j, \ldots, X_n$ of independent and identically distributed random variables. When one knows the parametric family generating this set of independent data, this information may be used to construct testing and estimation methods tailored to the individual distribution. The variety of such techniques is so large that only those presuming approximately a Gaussian model will be reviewed in this volume, and in what follows, parametric analyses for comparison purposes will be taken to assume approximate Gaussian distributions.

2.1.1 Estimation Using Averages

Practitioners often estimate the location of a distribution using the sample average

$$\bar{X} = \sum_{j=1}^{n} X_j / n. \tag{2.1}$$

If a new data set is created using an affine transformation $Y_j = a + bX_j$, then $\bar{Y} = a + b\bar{X}$, and the sample average is equivariant under affine transformations. For example, average temperature in degrees Fahrenheit $\bar{Y}$ may be calculated from average temperature in degrees Celsius $\bar{X}$ using $\bar{Y} = 32 + 1.8\bar{X}$, without needing access to the original measurements.

If these variables have a finite variance σ^2, then the central limit theorem (CLT) ensures that $\bar{X}$ is approximately $\mathcal{G}(\theta, \sigma^2/n)$; however, many common techniques designed for data with a Gaussian distribution require consequences of this distribution beyond the marginal distribution of the sample average.

2.1.2 One-Sample Testing for Gaussian Observations

To test the null hypothesis $\theta = \theta^0$ versus the alternative $\theta > \theta^0$, reject the null hypothesis if $\bar{X} > \theta^0 + z_\alpha \sigma/\sqrt{n}$. To test the null hypothesis $\theta = \theta^0$ versus the two-sided alternative $\theta \neq \theta^0$, reject the null hypothesis if $\bar{X} > \theta^0 + z_{\alpha/2}\sigma/\sqrt{n}$, or if $\bar{X} < \theta^0 - z_{\alpha/2}\sigma/\sqrt{n}$. If σ is not known, substitute the estimate $s = \sqrt{\sum_{j=1}^n (X_j - \bar{X})^2/(n-1)}$, and compare this quantity to the t distribution with $n-1$ degrees of freedom.

2.2 The Need for Distribution-Free Tests

Table 2.1 contains actual test levels for some tests of location parameters for four of the families described in §1.1.1. True levels were determined via simulation; a large number of samples were drawn from each of the distributions under the null hypothesis, the specified test statistic was calculated, the test of §2.1.2 was performed for each simulated data set, and the proportion of times the null hypothesis was rejected was tabulated. For now, restrict attention to the first line in each subtable, corresponding to the t-test. Null hypotheses in Table 2.1 are in terms of the distribution median. The t-test, however, is appropriate for hypotheses involving the expectation. In the Gaussian, Laplace, and uniform cases, the median coincides with the expectation, and so standard asymptotic theory justifies the use of the t-test. In the Cauchy example, as noted before, even though the distribution is symmetric, no expectation exists, and the t-test is inappropriate. However, generally, data analysts do not have sufficient information to distinguish the Cauchy example from the set of distributions having enough moments to justify the t-test, and so it is important to study the implications of such an inappropriate use of methodology.

For both sample sizes, observations from a Gaussian distribution give the targeted level, as expected. Observations from the Laplace distribution give a level close to the targeted level. Observations from the Cauchy distribution give a level much smaller than the targeted level, which is paradoxical, because one might expect heavy tails to make it anti-conservative. Figure 2.1 shows the density resulting from Studentizing the average of independent Cauchy variables. The resulting density is bimodal, with tails lighter than one would otherwise expect. This shows that larger values of the sample standard deviation in the denominator of the Studentized statistic act more strongly than larger values of components of the average in the numerator.

TABLE 2.1: True levels for the T Test, and Sign Test, and Exact Sign Test, nominal level 0.05

(a) Sample size 10, Two-Sided

	Gaussian	Cauchy	Laplace	Uniform
T	0.05028	0.01879	0.04098	0.05382
Approximate Sign	0.02127	0.02166	0.02165	0.02060
Exact Sign	0.02127	0.02166	0.02165	0.02060

(b) Sample size 17, Two-Sided

	Gaussian	Cauchy	Laplace	Uniform
T	0.05017	0.02003	0.04593	0.05247
Approximate Sign	0.01234	0.01299	0.01274	0.01310
Exact Sign	0.04847	0.04860	0.04871	0.04898

(c) Sample size 40, Two-Sided

	Gaussian	Cauchy	Laplace	Uniform
T	0.04938	0.02023	0.04722	0.05029
Approximate Sign	0.03915	0.03952	0.03892	0.03904
Exact Sign	0.03915	0.03952	0.03892	0.03904

In all cases above, the t-test succeeds in providing a test level not much larger than the target nominal level. On the other hand, in some cases the true level is significantly below that expected.

This effect decreases as sample level increases.

2.3 One-Sample Median Methods

For moderate sample sizes, then, the standard one-sample t-test fails to control test level as the distribution of summands changes. Techniques that avoid this problem are developed in this section. These methods apply in broad generality, including in cases when the expectation of the individual observations does not exist. Because of this, inference about the population median rather than the expectation is pursued. Recall that the median θ of random variable X_j is defined so that

$$P[X_j \geq \theta] \geq 1/2, \quad P[X_j \leq \theta] \geq 1/2. \tag{2.2}$$

Below, the term median refers to the population version, unless otherwise specified.

FIGURE 2.1: Density of Studentized Cauchy made Symmetric, Sample Size 10

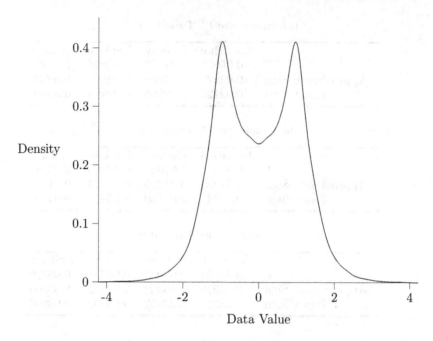

2.3.1 Estimates of the Population Median

An estimator smed$[X_1, \ldots, X_n]$ of the population median may be constructed by applying (2.2) to the empirical distribution of X_i, formed by putting point mass on each of the n values. In this case, with n odd, the median is the middle value, and, with n even, (2.2) fails to uniquely define the estimator. In this case, the estimator is conventionally defined to be the average of the middle two values. By this convention, with $X_{(1)}, \ldots, X_{(n)}$ the ordered values in the sample,

$$\text{smed}[X_1, \ldots, X_n] = \begin{cases} X_{((n+1)/2)} & \text{for } n \text{ odd} \\ (X_{(n/2)} + X_{(n/2+1)})/2 & \text{for } n \text{ even} \end{cases}. \quad (2.3)$$

Alternatively, one might define the sample median to minimize the sum of distances from the median:

$$\text{smed}[X_1, \ldots, X_n] = \underset{\eta}{\text{argmin}} \sum_{i=1}^n |X_i - \eta|; \quad (2.4)$$

that is, the estimate minimizes the sum of distances from data points to the potential median value, with distance measured by the sum of absolute values. This definition (2.4) exactly coincides with the earlier definition (2.3) for n odd, shares in the earlier definition's lack of uniqueness for even sample

One-Sample Median Methods

sizes, and typically shares the opposite resolution (averaging the middle two observations) of this non-uniqueness. In contrast, the sample mean $\bar{X}$ of (2.1) satisfies

$$\bar{X} = \underset{\eta}{\operatorname{argmin}} \sum_{i=1}^{n} |X_i - \eta|^2. \tag{2.5}$$

Under certain circumstances, the sample median is approximately Gaussian. Central limit theorems for the sample median generally require only that the density of the raw observations be positive in a neighborhood of the population median.

In §2.1.1 it was claimed that the sample average is equivariant for affine transformation. A stronger property holds for medians; if $Y_i = h(X_i)$, for h monotonic, then for n odd, smed $[Y_1,\ldots,Y_n] = h(\text{smed}\,[X_1,\ldots,X_n])$. For n even, this is approximately true, except that the averaging of the middle two observations undermines exact equivariance for non-affine transformations.

Both (2.4) and (2.5) are special cases of an estimator defined by

$$\underset{\eta}{\operatorname{argmin}} \sum_{i=1}^{n} \varrho(X_i - \eta), \tag{2.6}$$

for some convex function ϱ; the sample mean uses $\varrho(z) = z^2/2$ and the sample median uses $\varrho(z) = |z|$. Huber (1964) suggests an alternative estimator combining the behavior of the mean and median, by taking ρ quadratic for small values, and continuing linearly for larger values, thus balancing increased efficiency of the mean and the smaller dependence on outliers of the median; he suggests

$$\varrho(z) = \begin{cases} z^2/2 & \text{for } |z| < k \\ k|z| - k^2/2 & \text{for } |z| \geq k \end{cases}, \tag{2.7}$$

and recommends a value of the tuning parameter k between 1 and 2.

2.3.2 Hypothesis Tests Concerning the Population Median

Techniques in this section can be traced to Arbuthnott (1712), as described in the example below. Fisher (1930) treats this test as too obvious to require comment.

Consider independent identically distributed random variables X_i for $i = 1,\ldots,n$. To test whether a putative median value θ^0 is the true value, define new random variables

$$Y_j = \begin{cases} 1 & \text{i } X_j - \theta^0 \leq 0 \\ 0 & \text{i } X_j - \theta^0 > 0 \end{cases}. \tag{2.8}$$

Then under $H_0 : \theta = \theta^0$, $Y_j \sim \text{Bin}(1/2, 1)$. This logic only works if

$$\mathrm{P}\left[X_j = \theta^0\right] = 0; \tag{2.9}$$

assume this. It is usually easier to assess this continuity assumption than it is for distributional assumptions. Then the median inference problem reduces to one of binomial testing. Let $T(\theta^0) = \sum_{j=1}^{n} Y_j$ be the number of observations less than or equal to θ^0. Pick t_l and t_u so that $\sum_{j=t_l}^{t_u-1}(1/2)^n \binom{n}{j} \geq 1 - \alpha$. One might choose t_l and t_u symmetrically, so that t_l is the largest value such that

$$\sum_{j=0}^{t_l-1}(1/2)^n \binom{n}{j} \leq \alpha/2. \qquad (2.10)$$

That is, t_l is that potential value for T such that not more than $\alpha/2$ probability sits below it. The largest such t_l has probability at least $1 - \alpha/2$ equal to or larger than it, and at least $\alpha/2$ equal to or smaller than it; hence t_l is the $\alpha/2$ quantile of the $\mathfrak{B}\text{in}(n, 1/2)$ distribution. Generally, the inequality in (2.10) is strict; that is, $\leq$ is actually $<$. For combinations of n and α for which this inequality holds with equality, the quantile is not uniquely defined, and take the quantile to be the lowest candidate. Symmetrically, one might choose the smallest t_u so that

$$\sum_{j=t_u}^{n}(1/2)^n \binom{n}{j} \leq \alpha/2; \qquad (2.11)$$

$n+1-t_u$ is the $\alpha/2$ quantile of the $\mathfrak{B}\text{in}(n, 1/2)$ distribution, with the opposite convention used in case the quantile is not uniquely defined.

Then, reject the null hypothesis if $T \leq t_L^\circ$ or $T \geq t_U^\circ$ for $t_L^\circ = t_l - 1$ and $t_U^\circ = t_u$. This test is called the (exact) sign test, or the binomial test (Higgins, 2004). An approximate version of the sign test might be created by selecting critical values from the Gaussian approximation to the distribution of $T(\theta^0)$.

Again, direct attention to Table 2.1. Both variants of the sign test succeed in keeping the test level no larger than the nominal value. However, the sign test variants, because of the discreteness of the binomial distribution, in some cases achieve levels much smaller than the nominal target. Subtable (a), for sample size 10, is the most extreme example of this; subtable (b), for sample size 17, represents the smallest reduction in actual sample size, and subtable (c), for sample size 40, is intermediate. Note further that, while the asymptotic sign test, based on the Gaussian approximation, is not identical to the exact version, for subtables (a) and (c) the levels coincide exactly, since for all simulated data sets, the p-values either exceed 0.05 or fail to exceed 0.05 for both tests. Subtable (b) exhibits a case in which for one data value, the exact and approximate sign tests disagree on whether p-values exceed 0.05.

Table 2.2 presents characteristics of the exact two-sided binomial test of the null hypothesis that the probability of success is half, with level $\alpha = 0.05$, applied to small samples. In this case, the two-sided p-value is obtained by doubling the one-sided p value.

For small samples ($n < 6$), the smallest one-sided p-value, $1/2^n$, is greater than .025, and the null hypothesis is never rejected. Such small samples are omitted from Table 2.2. This table consists of two subtables side by side, for

TABLE 2.2: Exact levels and exact and asymptotic lower critical values for symmetric two-sided binomial tests of nominal level 0.05

n	Critical Value $t_l - 1$ Exact	Critical Value $t_l - 1$ Asymptotic	Exact Levels	n	Critical Value $t_l - 1$ Exact	Critical Value $t_l - 1$ Asymptotic	Exact Levels
6	0	0	0.0313	24	6	6	0.0227
7	0	0	0.0156	25	7	7	0.0433
8	0	0	0.0078	26	7	7	0.0290
9	1	1	0.0391	27	7	7	0.0192
10	1	1	0.0215	28	8	8	0.0357
11	1	1	0.0117	29	8	8	0.0241
12	2	2	0.0386	30	9	9	0.0428
13	2	2	0.0225	31	9	9	0.0294
14	2	2	0.0129	32	9	9	0.0201
15	3	3	0.0352	33	10	10	0.0351
16	3	3	0.0213	34	10	10	0.0243
17	4	3	0.0490	35	11	11	0.0410
18	4	4	0.0309	36	11	11	0.0288
19	4	4	0.0192	37	12	12	0.0470
20	5	5	0.0414	38	12	12	0.0336
21	5	5	0.0266	39	12	12	0.0237
22	5	5	0.0169	40	13	13	0.0385
23	6	6	0.0347	41	13	13	0.0275

$n \leq 23$, and for $n > 23$. The first column of each subtable is sample size. The second is $t_l - 1$ from (2.10). The third is the value taken from performing the same operation on the Gaussian approximation; that is, it is the largest a such that

$$\Phi((t_l - 1 - n/2 + 0.5)/(0.5\sqrt{n})) \leq \alpha/2. \qquad (2.12)$$

The fourth is the observed test level; that is, it is double the right side of (2.10). Observations here agree with those from Table 2.1; for sample size 10, the level of the binomial test is severely too small, for sample size 17, the binomial test has close to the optimal level, and for sample size 40, the level for the binomial test is moderately too small.

A complication (or, to an optimist, an opportunity for improved approximation) arises when approximating a discrete distribution by a continuous distribution. Consider the case with $n = 10$, exhibited in Figure 2.2. Bar areas represent the probability under the null hypothesis of observing the number of successes. Table 2.2 indicates that the one-sided test of level 0.05 rejects the null hypothesis for $W \leq 1$. The actual test size is 0.0215, which is graphically represented as the sum of the areas in the bar centered at 1, and the very small area of the neighboring bar centered at 0. Expression (2.12) approximates the sum of these two bar areas by the area under the dotted curve, representing the Gaussian density with the appropriate expectation $n/2 = 5$ and standard deviation $\sqrt{n}/2 = 1.58$. In order to align the areas of the bars most closely with the area under the curve, the Gaussian area should be taken to extend to the upper end of the bar containing 1; that is, evaluate the Gaussian distribution function at 1.5, explaining the 0.5 in (2.12). More generally, for a discrete distribution with potential values Δ units apart, the ordinate is shifted by $\Delta/2$ before applying a Gaussian approximation; this adjustment is called a correction for continuity.

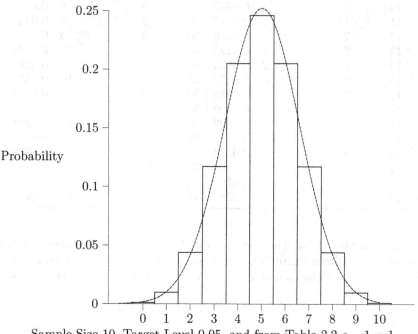

FIGURE 2.2: Approximate Probability Calculation for Sign Test
Sample Size 10, Target Level 0.05, and from Table 2.2 $a - 1 = 1$

The power of the sign test is determined by $P_{\theta^A}\left[X_j \leq \theta^0\right]$ for values of $\theta^A \neq \theta^0$. Since $\theta^A > \theta^0$ if $P_{\theta^A}\left[X_j \leq \theta^0\right] < 1/2$, alternatives $\theta^A > \theta^0$ correspond to one sided alternatives $P\left[Y_j = 1\right] < 1/2$.

If θ^0 is the true population median of the X_j, and if there exists a set of form $(\theta^0 - \epsilon, \theta^0 + \epsilon)$, with $\epsilon > 0$, such that $P\left[X_j \in (\theta^0 - \epsilon, \theta^0 + \epsilon)\right] = 0$, then any other θ in this set is also a population median for X_j, and hence the test will have power against such alternatives no larger than the test level. Such occurrences are rare.

Table 2.3 represents powers for these various tests for various sample levels. The alternative is chosen to make the t-test have power approximately .80 for the Gaussian and Laplace distributions, using (1.9). In this case both σ_0 and σ_A for the Gaussian and Laplace distributions are $1/\sqrt{n}$. Formula (1.9) is inappropriate for the Cauchy distribution, since in this case $\bar{X}$ does not have

One-Sample Median Methods

TABLE 2.3: Power for the T Test, Sign Test, and Exact Sign Test, nominal level 0.05

(a) Sample size 10, Two-Sided

	Gaussian	Cauchy	Laplace
T	0.70593	0.14345	0.73700
Approximate Sign	0.41772	0.20506	0.57222
Exact Sign	0.41772	0.20506	0.57222

(b) Sample size 17, Two-Sided

	Gaussian	Cauchy	Laplace
T	0.74886	0.10946	0.76456
Approximate Sign	0.35747	0.17954	0.58984
Exact Sign	0.57893	0.35759	0.79011

(c) Sample size 40, Two-Sided

	Gaussian	Cauchy	Laplace
T	0.78152	0.06307	0.78562
Approximate Sign	0.55462	0.35561	0.84331
Exact Sign	0.55462	0.35561	0.84331

a distribution that is approximately Gaussian. For the Cauchy distribution, the same alternative as for the Gaussian and Laplace distributions is used.

Results in Table 2.3 show that for a sample size for which the sign test level approximates the nominal level ($n = 17$), use of the sign test for Gaussian data results in a moderate loss in power relative to the t-test, while use of the sign test results in a moderate gain in power for Laplace observations, and in a substantial gain in power for Cauchy observations.

Example 2.3.1 *An early (and very simple) application of this test was to test whether the proportion of boys born in a given year is the same as the proportion of girls born that year (Arbuthnott, 1712). Number of births was determined for a period of 82 years. Let X_j represent the number of births of boys, minus the number of births of girls, in year j. The parameter θ represents the median amount by which the number of girls exceeds the number of boys; its null value is 0. Let Y_j take the value 0 for years in which more girls than boys are born, and 1 otherwise. Note that in this case, (2.9) is violated, but $P[X_j = 0]$ is small, and this violation is not important. Test at level 0.05.*

Values in (2.10) and (2.11) are $t_l = 32$ and $t_u = 51$, obtained as the 0.025 and 0.975 quantiles of the binomial distribution with 82 trials and success probability .5. Reject the null hypothesis if $T < 32$ or if $T \geq 51$. (The asymmetry in the treatment of the lower and upper critical values is intentional, and is the result of the asymmetry in the definition of the distribution function for discrete variables.)

In each of these years $X_j < 0$, and so $Y_j = 1$, and $T = 82$. Reject the null hypothesis of equal proportion of births. The original analysis of this data presented what is now considered the p-value; the one sided value of (1.14) is trivially $P[T \geq= 82] = (1/2)^{82}$, which is tiny. The two-sided p-value of (1.15) is $2 \times (1/2)^{82} = (1/2)^{81}$, which is still tiny.

2.3.3 Confidence Intervals for the Median

Apply the test inversion approach of §1.2 to the sign test that rejects H_0: $\theta = \theta^0$ if fewer than t_l or at least t_u data points are less than or equal to θ^0. Let $X_{(\cdot)}$ referring to the data values after ordering. When $\theta^0 \leq X_{(1)}$, then $T(\theta^0) = 0$. For $\theta^0 \in (X_{(1)}, X_{(2)}]$, $T(\theta^0) = 1$. For $\theta^0 \in (X_{(2)}, X_{(3)}]$, $T(\theta^0) = 2$. In each case, the (at the beginning of the interval and the] at the end of the interval arise from (2.8), because observations that are exactly equal to θ^0 are coded as one. Hence the test rejects H_0 if $\theta^0 \leq X_{(t_l)}$ or $\theta^0 > X_{(t_u)}$, and, for any θ^0,

$$P_{\theta^0}\left[X_{(t_l)} < \theta^0 \leq X_{(t_u)}\right] \geq 1 - \alpha.$$

This relation leads to the confidence interval $(X_{(t_l)}, X_{(t_u)}]$. However, since the data have a continuous distribution, then $X_{(t_u)}$ also has a continuous distribution, and

$$P_{\theta^0}\left[\theta^0 = X_{(t_u)}\right] = 0$$

for any θ^0. Hence $P_{\theta^0}\left[X_{(t_l)} < \theta^0 < X_{(t_u)}\right] \geq 1 - \alpha$, and one might exclude the upper end point, to obtain the interval $(X_{(t_l)}, X_{(t_u)})$.

Example 2.3.2 *Consider data from*

http://lib.stat.cmu.edu/datasets/Arsenic

from a pilot study on the uptake of arsenic from drinking water. Column six of this file gives arsenic concentrations in toenail clippings, in parts per million. The link above is to a Word file; the file

http://stat.rutgers.edu/home/kolassa/Data/arsenic.dat

contains a plain text version. Sorted nail arsenic values are

0.073, 0.080, 0.099, 0.105, 0.118, 0.118, 0.119, 0.135, 0.141, 0.158, 0.175, 0.269, 0.275, 0.277, 0.310, 0.358, 0.433, 0.517, 0.832, 0.851, 2.252.

We construct a confidence interval for the (natural) log of toenail arsenic. The sign test statistic has a $\mathfrak{B}in(21, .5)$ distribution under the null hypothesis. We choose the largest t_l such that $P_0[T < t_l] \le \alpha/2$. The first few terms in (2.10) are

$$4.76 \times 10^{-7}, 1.00 \times 10^{-5}, 1.00 \times 10^{-4}, 6.34 \times 10^{-4}, 2.85 \times 10^{-3},$$
$$9.70 \times 10^{-3}, 2.59 \times 10^{-2}, 5.54 \times 10^{-2}, 9.70 \times 10^{-2}, 1.40 \times 10^{-1},$$

and cumulative probabilities are

$$4.77 \times 10^{-7}, 1.05 \times 10^{-5}, 1.11 \times 10^{-4}, 7.45 \times 10^{-4}, 3.60 \times 10^{-3},$$
$$1.33 \times 10^{-2}, 3.91 \times 10^{-2}, 9.46 \times 10^{-2}, 1.92 \times 10^{-1}, 3.32 \times 10^{-1}.$$

The largest of these cumulative sums smaller than 0.025 is the sixth, corresponding to $T < 6$. Hence $t_l = 6$. Similarly, $t_u = 16$. Reject the null hypothesis that the mean is 0.26 if $T < 6$ or if $T \ge 16$. Since 11 of the observations are greater than the null median 0.26, $T = 11$. Do not reject the null hypothesis.

Alternatively, one might calculate a p-value. Using (1.15), the p-value is $2\min(P_0[T \ge 11], P_0[T \le 11]) = 1$.

Furthermore, the confidence interval for the median is $(X_{(6)}, X_{(16)}) = (0.118, 0.358)$.

The values t_l and t_u may be calculated in R by

```
a<-qbinom(0.025,21,.5); b<-21+1-qbinom(0.025,21,.5)
```

and the ensemble of calculations might also have been performed in R using

```
arsenic<-as.data.frame(scan('arsenic.dat',
    what=list(age=0,sex=0,drink=0,cook=0,water=0,nails=0)))
library(BSDA)#Gives sign test.
SIGN.test(arsenic$nails,md=0.26)#Argument md gives null hyp.
```

Graphical construction of a confidence interval for the median is calculated by

```
library(NonparametricHeuristic)
invertsigntest(log(arsenic$nails),maint="Log Nail Arsenic")
```

and is given in Figure 2.3. Instructions for installing this last library are given in the introduction, and in Appendix B.

Figure 2.3 exhibits construction of the confidence interval in the previous example; I apply these techniques on the log scale. The confidence interval is the set of log medians that yield a test statistic for which the null hypothesis is not rejected. Values of the statistic for which the null hypothesis is not rejected are between the horizontal lines; log medians in the confidence intervals are values of the test statistic within this region.

FIGURE 2.3: Construction of CI for Log Nail Arsenic Location

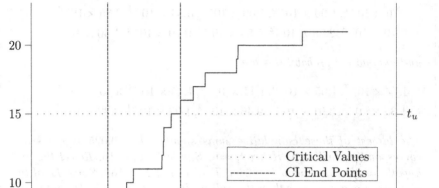

In this construction, <u>order statistics</u> (that is, the ordered values) are first plotted on the horizontal axis, with the place in the ordered data set on the vertical axis. These points are represented by the points in Figure 2.3 where the step function transitions from vertical to horizontal, as one moves from lower left to upper right. Next, draw horizontal lines at the values t_l and t_u, given by (2.10) and (2.11) respectively. Finally, draw vertical lines through the data points that these horizontal lines hit.

For this particular example, the exact one-sided binomial test of level 0.025 rejects the null hypothesis that the event probability is half if the sum of event indicators is 0, 1, 2, 3, 4, or 5; $t_l = 6$. For Y_j of (2.8), the sum is less than 6 for all θ to the left of the point marked $X_{(t_l)}$. Similarly, the one-sided level 0.025 test in the other direction rejects the null hypothesis if the sum of event indicators is at least $t_u = 16$. The sum of the Y_j exceeds 15 for θ to the right of the point marked $X_{(t_u)}$.

By symmetry, one might expect $t_l = n - t_u$, but this is not the case. The asymmetry in definitions (2.10) and (2.11) arises because construction of the confidence interval requires counting not the data points, but the $n-1$ spaces

One-Sample Median Methods 27

between them, plus the regions below the minimum and above the maximum, for a total of $n + 1$ ranges. Then $t_l = n + 1 - t_u$.

This interval is not of the usual form $\hat{\theta} \pm 2\hat{\sigma}$, for $\hat{\sigma}$ with a factor of $1/\sqrt{n}$. Cramér (1946, pp. 368f.) shows that if $X_1, \ldots, X_n$ is a set of independent random variables, each having density f, then Var $[\text{smed}[X_1, \ldots, X_n]] \approx 1/(4f(\theta)^2 n)$. Chapter 8 investigates estimation of this density; this estimate can be used to estimate the median variance, but density estimation is harder than the earlier confidence interval rule.

2.3.4 Inference for Other Quantiles

The quantile θ corresponding to probability γ is defined by $P_\theta[X_j \leq \theta] = \gamma$. Suppose that θ is quantile $\gamma \in (0, 1)$ of distribution of independent and identically distributed continuous random variables $X_1, \ldots, X_n$. Then one can produce a generalized sign test. Define the null and alternative hypotheses $H_0 : \theta = \theta^0$ and $H_A : \theta \neq \theta^0$. As before, $T(\theta)$ is the number of observations smaller than or equal to θ. For the true value θ of the quantile, $T \sim \text{Bin}(n, \gamma)$. Choose t_l and t_u so that $\sum_{j=t_l}^{t_u-1} \gamma^j (1-\gamma)^{n-j} \binom{n}{j} \geq 1 - \alpha$. Often, one chooses the largest t_l and smallest t_u so that

$$\sum_{j=0}^{t_l-1} \gamma^j (1-\gamma)^{n-j} \binom{n}{j} < \alpha/2, \quad \sum_{j=t_u}^{n} \gamma^j (1-\gamma)^{n-j} \binom{n}{j} < \alpha/2 \quad (2.13)$$

this t_l is $\alpha/2$ quantile of the $\text{Bin}(n, \gamma)$ distribution, and $n + 1 - t_u$ is the $\alpha/2$ quantile of the $\text{Bin}(n, 1 - \gamma)$ distribution. Hence $t_l \approx n\gamma - \sqrt{n\gamma(1-\gamma)}z_{\alpha/2}$ and $t_u \approx n\gamma + \sqrt{n\gamma(1-\gamma)}z_{\alpha/2}$. One then rejects H_0 if $T < t_l$ or $T \geq t_u$.

This test is then inverted to obtain $(X_{(t_l)}, X_{(t_u)})$ as the confidence interval for θ. Note that the confidence level is conservative: $P[X_{(t_l)} \leq \theta \leq X_{(t_u)}] = 1 - P[X_{(t_l)} \geq \theta] - P[\theta \geq X_{(t_u)}] \geq 1 - \alpha$. For any given θ, the inequality is generally strict.

Example 2.3.3 *Test the null hypothesis that the upper quartile (that is, the 0.75 quantile) of the arsenic nail data from Example 2.3.2 is the reference value 0.26, and give a confidence interval for this quantile. The analysis is the same as before, except that t_l and t_u are different. We determine t_l and t_u in (2.13). Direct calculation, or using the R commands*

```
a<-qbinom(0.025,21,.75);b<-21+1-qbinom(0.025,21,1-0.75)
```

shows $t_l = 12$ and $t_u = 20$. Since $T = 10 < t_l$, reject the null hypothesis that the upper quartile is 0.26. Furthermore, the confidence interval is the region between the twelfth and twentieth ordered values, $(X_{(12)}, X_{(20)}) = (0.269, 0.851)$. With data present in the R workspace, one calculates a confidence interval as

```
sort(arsenic$nails)[c(a,b)]
```

and the p-value as

```
tt<-10
2*min(c(pbinom(tt,21,.75), pbinom(21+1-tt,21,1-.75)))
```

to give 0.0128.

Dependence of the test statistic $T(\theta)$ on θ is relatively simple. Later inversions of more complicated statistics will make use of the simplifying device of first, shifting all or part of the data by subtracting θ, and then testing the null hypothesis that the location parameter for this shifted variable is zero.

2.4 Comparing Tests

For fixed level, alternative, and power, the test with a smaller sample size is better. Consider two families of one-sided tests, indexed by sample size, using statistics T_1 and T_2, both with test level α, and determine the sample sizes required to give power $1 - \beta$, for the same alternative. Compare the tests by taking the ratio of these two sample sizes. The ratio is called relative efficiency; the notation dates back at least as far as Noether (1950), citing Pitman (1948).

Let $t^\circ_{j,n}$ represent the critical value for test j based on n observations; that is, the test based on statistic T_j and using n observations, rejects the null hypothesis if $T_j \geq t^\circ_{j,n}$. Hence $t^\circ_{j,n}$ satisfies $P_{\theta^0}\left[T_j \geq t^\circ_{j,n}\right] = \alpha$. Let $\varpi_{j,n}(\theta^A)$ represent the power for test T_j using n observations, under the alternative θ^A:

$$\varpi_{j,n}(\theta^A) = P_{\theta^A}\left[T_j \geq t^\circ_{j,n}\right].$$

Assume that

$$\left.\begin{array}{l} \varpi_{j,n}(\theta^A) \text{ is continuous and increasing in } \theta^A \text{ for all } j, n, \\ \lim_{\theta^A \to \infty} \varpi_{j,n}(\theta^A) = 1, \\ \lim_{n \to \infty} \varpi_{j,n}(\theta^A) = 1 \text{ for all } \theta^A > \theta^0. \end{array}\right\} \quad (2.14)$$

Two tests, tests 1 and 2, involving hypotheses about a parameter θ, taking the value θ^0 under the null hypothesis, and with a simple alternative hypothesis of form $\{\theta^A\}$, for some $\theta^A > \theta^0$, with similar level and power, will be compared. Pick a test level α and a power $1 - \beta$, and the sample size n_1 for test 1. The power and level conditions on T_1 imply a value for θ^A under the alternative hypothesis; that is, θ^A solves $P_{\theta^A}\left[T_1 \geq t^\circ_{1,n_1}\right] = 1 - \beta$. Note that θ^A is a function of n_1, α, and β. Under conditions (2.14), one can determine the minimal value of n_2 so that test 2 has power at least $1 - \beta$, under the alternative given by θ^A. Report n_1/n_2 as the relative efficiency of test 2 to test 1; this depends on n_1, α, and β.

Define the asymptotic relative efficiency $\text{ARE}_{\alpha,\beta}[T_1, T_2]$ as

$$\lim_{n_1 \to \infty} n_1/n_2,$$

when this limit exists. Considering this quantity removes dependence on n_1.

This measure comparing efficiencies of two tests takes on a particularly easy form in a special, yet common, case, in which both statistics are asymptotically Gaussian. In this case, the relative efficiency can be approximated in terms of standard deviations and derivatives of means under alternative hypotheses. General approximations for sample size, power, and effect sizes are investigated first; these are applied to relative efficiency later.

2.4.1 Power, Sample Size, and Effect Size

This subsection presents formulas for power, sample size, and effect size, that may be used for efficiency comparisons, but are also useful on their own. Gaussian approximations earlier in this chapter often applied a continuity correction; this correction will not be applied for large-sample power and sample size calculations, as the effect of this correction quickly becomes negligible as the sample size increases. Without loss of generality, take $\theta^0 = 0$.

2.4.1.1 Power

Consider test statistics satisfying

$$T_j \sim \mathfrak{G}(\mu_j(\theta), \varsigma_j^2(\theta)), \text{ for } \varsigma_j(\theta) > 0, \mu_j(\theta) \text{ increasing in } \theta. \quad (2.15)$$

The Gaussian distribution in (2.15) does not need to hold exactly; holding approximately is sufficient. In this case, one can find the critical values for the two tests, t_{j,n_j}°, such that $\text{P}_0\left[T_j \geq t_{j,n_j}^\circ\right] = \alpha$. Since $(T_j - \mu_j(0))/\varsigma_j(0)$ is approximately standard Gaussian under the null hypothesis, then

$$\alpha = \text{P}_0\left[(T_j - \mu_j(0))/\varsigma_j(0) \geq z_\alpha\right] = \text{P}_0\left[T_j \geq \mu_j(0) + \varsigma_j(0)z_\alpha\right].$$

Hence
$$t_{j,n_j}^\circ = \mu_j(0) + \varsigma_j(0)z_\alpha. \quad (2.16)$$

The power for test j is approximately

$$\begin{aligned}\varpi_{j,n_j}(\theta^A) &\approx \text{P}_{\theta^A}\left[T_j \geq \mu_j(0) + \varsigma_j(0)z_\alpha\right] \\ &= 1 - \Phi\left(\left[\mu_j(0) + \varsigma_j(0)z_\alpha - \mu_j(\theta^A)\right]/\varsigma_j(\theta^A)\right). \quad (2.17)\end{aligned}$$

Often the variance of the test statistic changes slowly as one moves away from the null hypothesis; in this case, the power for test j is approximately

$$\varpi_{j,n_j}(\theta^A) \approx 1 - \Phi\left(\left[\mu_j(0) - \mu_j(\theta^A)\right]/\varsigma_j(0) + z_\alpha\right). \quad (2.18)$$

2.4.1.2 Sample and Effect Sizes

When the test statistic variance decreases in a regular way with sample size, one can invert the power relationship to determine the sample size needed for a given power and effect size. Consider tests satisfying, in addition to (2.15),

$$\varsigma_j^2(\theta) = \sigma_j^2(\theta)/n_j. \tag{2.19}$$

Then

$$\varpi_{j,n_j}(\theta^A) = 1 - \Phi\left(\sqrt{n_j}\left[\mu_j(0) + \frac{\sigma_j(0)z_\alpha}{\sqrt{n_j}} - \mu_j(\theta^A)\right]/\sigma_j(\theta^A)\right). \tag{2.20}$$

As sample sizes increase, power increases for a fixed alternative, and calculations will consider a class of alternatives moving towards the null. Calculations below will consider behavior of the expectation of the test statistic near the null hypothesis (which is taken as $\theta = 0$). Suppose that

$$\mu_j(\theta), \sigma_j(\theta) \text{ are differentiable on some set } \theta \in (-\epsilon, \epsilon). \tag{2.21}$$

(These conditions are somewhat simpler than considered by Noether (1950); in particular, note that (2.15), (2.19), and (2.21) together are not enough to demonstrate the second condition of (2.14).) Without loss of generality, continue to take $\theta^0 = 0$. In this case, critical values for the two tests are given in (2.16). The power expression (2.17) may be simplified by approximating the variances at the alternative hypothesis by quantities at the null. For large n_j, alternatives with power less than 1 will have alternative hypotheses near the null, and so $\sigma_j(\theta^A) \approx \sigma_j(\theta^0)$. Hence

$$\begin{aligned}\varpi_{j,n_j}(\theta^A) &\approx 1 - \Phi\left(\sqrt{n_j}\left[\frac{\mu_j(0) - \mu_j(\theta^A)}{\sigma_j(0)}\right] + z_\alpha\right) \\ &= \Phi\left(\sqrt{n_j}\left[\frac{\mu_j(\theta^A) - \mu_j(0)}{\sigma_j(0)}\right] - z_\alpha\right).\end{aligned} \tag{2.22}$$

This expression for approximate power may be solved for sample size, by noting that if

$$\varpi_{j,n_j}(\theta^A) = 1 - \beta, \tag{2.23}$$

then $\varpi_{j,n_j}(\theta^A) = \Phi(z_\beta)$, and (2.22) holds if $\sqrt{n_j}\left[\frac{\mu_j(\theta^A)-\mu_j(0)}{\sigma_j(0)}\right] - z_\alpha = z_\beta$, or

$$n_j = \sigma_j(0)^2(z_\alpha + z_\beta)^2/(\mu_j(\theta^A) - \mu_j(0))^2. \tag{2.24}$$

Common values for α and β are 0.025 and 0.2, giving upper Gaussian quantiles of $z_\alpha = 1.96$ and $z_\beta = 0.84$. Recall that z with a subscript strictly between 0 and 1 indicates that value for which a standard Gaussian random variable has that probability above it.

It may be of use in practice, and will be essential in the efficiency calculations below, to approximate which member of the alternative hypothesis

Comparing Tests

corresponds with a test of a given power, with sample size held fixed. Solving (2.24) exactly for θ^A is difficult, since the function μ is generally non-linear. Approximating this function using a one-term Taylor approximation,

$$\mu_j(\theta^A) \approx \mu_j(0) + \mu'_j(0)\theta^A.$$

(Contrast this with approximation of the alternative standard deviation by the null standard deviation, as in the transition from (2.20) to (2.22). Approximation by the leading term alone cannot be applied to the expectation $\mu_j(\theta)$, since it would remove all of the effect of the difference between null and alternative.) The power for test j is approximately

$$1 - \Phi\left((\sigma_j(0)z_\alpha - \sqrt{n_j}\mu'_j(0)\theta^A)/\sigma_j(0)\right) = 1 - \Phi\left(z_\alpha - \sqrt{n_j}e_j\theta^A\right)$$

for

$$e_j = \mu'_j(0)/\sigma(0).$$

The quantity e_j is called the efficacy of test j. Setting this power to $1 - \beta$, $z_\alpha - \sqrt{n_j}e_j\theta^A = z_{1-\beta}$. Solving this equation for θ^A,

$$\theta^A \approx (z_\alpha + z_\beta)/\sqrt{n_j}e_j, \quad (2.25)$$

verifying the requirement that θ^A get close to zero. This expression can be used to approximate an effect size needed to obtain a certain power with a certain sample size and test level, and will be used in the context of asymptotic relative efficiency.

2.4.2 Efficiency Calculations

Equating the alternative hypothesis parameter values (2.25) corresponding to power $1 - \beta$, then $(z_\alpha - z_{1-\beta})/\sqrt{n_1}e_1 = (z_\alpha - z_{1-\beta})/\sqrt{n_2}e_2$, or

$$\text{ARE}_{\alpha,\beta}[T_1, T_2] = n_2/n_1 = e_1^2/e_2^2.$$

Note that this relative efficiency doesn't depend on α or β, or on n_1. As an example, suppose $X_1, \ldots, X_n$ are independent observations from a symmetric distribution with finite variance ρ^2 and mean θ. Then θ is also the median of these observations. Compare tests T_1, the t-test, and T_2, the sign test. Then T_1 has a distribution depending on the distribution of X_j, and T_2 has a binomial distribution. Note that T_1 has approximately a standard Gaussian distribution for large n_1. That is, $T_1 \sim \mathfrak{G}(\theta/\rho, 1/n_1)$, and

$$\mu'_1(0) = 1/\rho, \quad \sigma_1(0) = 1, \text{ and } e_1 = 1/\rho. \quad (2.26)$$

On the other hand, $T_2 \sim \mathfrak{G}(\mu_2(\theta), \sigma_2(\theta)^2/n_2)$ for

$$\mu_2(\theta) = F(\theta), \quad \sigma_2(\theta) = \sqrt{F(\theta)(1 - F(\theta))}. \quad (2.27)$$

Hence $\mu'_2(0) = f(0)$, and $\sigma_2(0) = 1/2$, and $e_2 = 2f(0)$.

TABLE 2.4: Empirical powers for one-sample location tests with sample size ratios indicated by asymptotic relative efficiency

Larger sample size	t test	sign test
20	0.5623	0.2241
100	0.5647	0.3897
1000	0.5594	0.5040
10000	0.5621	0.5438

The asymptotic relative efficiencies of these statistics depends on the distribution that generates the data. If data come from $\mathfrak{G}(\theta, \rho^2)$, then $\mu_2'(0) = 1/(\sqrt{2\pi}\rho)$, $\sigma_2(0) = 1/2$, and $e_2 = \sqrt{2/\pi}/\rho$. Then $n_1/n_2 \approx (2/\sqrt{2\pi})^2 = 2/\pi$. Hence, as expected, the t-test is more powerful; the sign test requires more than 50% more observations to obtain the same power against the same alternative, for large samples.

If the data come from a Laplace distribution, then $\rho = 1$, since the Laplace distribution has variance 1. Substituting into (2.26), $\mu_1'(0) = 1$, $\sigma_1(0) = 1$, and $e_1 = 1$. Also $\mu_2'(0) = 1/\sqrt{2}$, $\sigma_2(0) = 1/2$, and $e_2 = \sqrt{2}$. Hence $n_1/n_2 \approx (\sqrt{2})^2 = 2$; in this case, the sign test is more powerful, requiring roughly half the sample size as does the t-test.

Table 2.4 contains results of a simulation to check actual powers that the asymptotic relative efficiency calculations show should be approximately the same. The table shows powers of the level 0.05 two-sided t and exact sign tests for Laplace data sets, of size n_1 and $n_2 = n_1/2$ respectively, shifted to have expectation $3/n_1$. Data sets need to be quite large in order for sample sizes in the ratio of the asymptotic relative efficiency to give equal power.

Now suppose these data come from a Cauchy distribution shifted to have point of symmetry θ. In this case, the expectation of the distribution does not exist, the standard deviation ρ is infinite, and the distribution is not approximately Gaussian even in large samples. In fact, the distribution of the the mean of Cauchy random variables is again a Cauchy random variable, with no change in the spread of the distribution. Plugging into the definition of efficacy, without worrying about regularity conditions, gives $\mu_1'(0) = 1$, $\sigma_1(0) = \infty$, and $e_1 = 0$. On the other hand, the quantities for the sign test are

$$\mu_2'(0) = \pi^{-1}, \quad \sigma_2(0) = 1/2, \text{ and } e_2 = 2/\pi.$$

Hence, for Cauchy responses, the efficiency of the sign test relative to the t-test is $n_1/n_2 \approx \infty$. This abuse of notation retains the interpretation that the sign test is infinitely more efficient for Cauchy observations.

Table 2.5 summarizes these calculations.

Comparing Tests 33

TABLE 2.5: Efficacies for one-sample location tests

		t-test	Sign test	Relative
Gaussian	$\mu'(0)$	$1/\rho$	$1/(\sqrt{2\pi}\rho)$	
	$\sigma(0)$	1	1/2	
	e	$1/\rho$	$\sqrt{2/\pi}/\rho$	$\sqrt{\pi/2}$
Laplace	$\mu'(0)$	1	$1/\sqrt{2}$	
	$\sigma(0)$	1	1/2	
	e	1	$\sqrt{2}$	$1/\sqrt{2}$
Cauchy	$\mu'(0)$	1	π^{-1}	
	$\sigma(0)$	∞	1/2	
	e	0	$2/\pi$	0

2.4.3 Examples of Power Calculations

The Gaussian approximations to power (2.17), to sample size (2.24), and to effect size (2.25), may be used to assist in planning an experiment.

Example 2.4.1 *In this example I calculate power for a sign test applied to 49 observations from a Gaussian distribution with unit variance. Suppose $X_1, \ldots, X_{49} \sim \mathfrak{G}(\theta, 1)$, with null hypothesis $\theta = 0$ and alternative hypothesis $\theta = 1/2$. The sign test statistic, divided by n, approximately satisfies (2.15) and (2.19), with μ and σ given by (2.27). Then $\mu_1(0) = .5$, $\sigma_1(0) = \sqrt{0.5 \times 0.5} = .5$, $\mu_1(0.5) = 0.691$, $\sigma_1(0.5) = \sqrt{.691 \times 0.309} = 0.462$, and power for a one-sided test of level 0.025, or a two-sided test of level 0.05, is approximated by (2.17): $1 - \Phi(7 \times (0.5 + 0.5 \times 1.96/7 - 0.691)/0.462) = 1 - \Phi(-0.772) = 0.780$. The null and alternative standard deviations are close enough to motivate the use of the simpler approximation (2.22), approximating power as*

$$1 - \Phi(7 \times (0.5 - 0.691)/0.5 + 1.96) = 1 - \Phi(-0.714) = 0.769.$$

If, instead, a test of power 0.85 were desired for alternative expectation 1/2, with a one-sided test of level 0.025, $z_\alpha = 1.96$, and $z_\beta = 1.036$. From (2.24), one needs at least

$$n = (0.462)^2(1.96 + 1.036)^2/(0.691 - 0.5)^2 = 52.52$$

observations; choose 53.

Finally, one might determine how large an effect one might detect using the original 49 observations with a test of level 0.025 and power 0.85. One could use $e = \sqrt{2/\pi} = 0.797$, from the box in Table 2.5 specific to the sign test and the Gaussian distribution. Expression (2.25) gives this number as $(1.96 + 1.036)/(7 \times 0.797) = 0.537$.

2.5 Distribution Function Estimation

Suppose one wishes to estimate a common distribution function of $X_1,\ldots,X_n$ independent variables. For x in the range of X_j, let $\hat{F}(x)$ be the number of data points less than or equal to x, divided by n. Since the observations are independent, $\hat{F}(x) \sim n^{-1}\text{Bin}(n, F(x))$. A confidence interval for $F(x)$ is

$$\hat{F}(x) \pm z_{\alpha/2}\sqrt{\hat{F}(x)(1-\hat{F}(x))/n}. \tag{2.28}$$

The above intervals will extend outside $[0,1]$, which is not reasonable; this can be circumvented by transforming the probability scale.

Figure 2.4 represents the bounds from (2.28), without any rescaling, to be discussed further in the next example. Confidence bounds in Figure 2.4 exhibit occurrences of larger estimates being associated with upper confidence bounds that are smaller (ex., in Figure 2.4, the region between the second-to-largest and the largest observations), and for the region with the cumulative distribution function estimated at zero or one (that is, the region below the smallest observed value, and the region above the largest observed value), confidence limits lie on top of the estimates, indicating no uncertainty. Both of these phenomena are unrealistic. The first phenomenon, that of non-monotonic confidence bounds, cannot be reliably avoided through rescaling; the second, with the upper confidence bounds outside the range of the data, can never be repaired through rescaling. A preferred solution is to substitute the intervals of Clopper and Pearson (1934), described in §1.2.2.1, to avoid all three of these problems (viz., bounds outside $(0,1)$, bounds ordered differently than the estimate, and bounds with zero variability). Such intervals are exhibited in Figure 2.5.

Finally, the confidence associated with these bounds is point-wise, and not simultaneous. That is, if (L_1, U_1) and (L_2, U_2) and are $1-\alpha$ confidence bounds associated with two ordinates x_1 and x_2, then $\text{P}\left[L_1 \le F(x_1) \le U_1\right] \ge 1 - \alpha$ and $\text{P}\left[L_2 \le F(x_2) \le U_2\right] \ge 1 - \alpha$, at least approximately, but the preceding argument does not bound $\text{P}\left[L_1 \le F(x_1) \le U_1 \text{ and } L_2 \le F(x_2) \le U_2\right]$ any higher than $1 - 2\alpha$.

Example 2.5.1 *Consider the arsenic data of Example 2.3.2. For every real x, one counts the number of data points less than this x. For any x less than the smallest value 0.073, this estimate is $\hat{F}(x) = 0$. For x greater than or equal to this smallest value and smaller than the next smallest value 0.080, the estimate is $\hat{F}(x) = 1/21$. This data set contains one duplicate value 0.118. For values below, but close to, 0.118 (for example, $x = 0.1179$), $\hat{F}(x) = 4/21$, since 21 of the observations are less than x. However, $\hat{F}(x) = 6/21$; the jump here is twice what it is at other data values, since there are two observations here. This esti-*

Exercises 35

> mate is sketched in both Figures 2.4 and 2.5, and may be constructed in R using ecdf(arsenic$nails), presuming the data of Example 2.3.2 is still present to R. The command ecdf(arsenic$nails) does not produce confidence intervals; use
>
> library(MultNonParam); ecdfcis(arsenic$nails,exact=FALSE)
>
> to add confidence bounds, and changing exact to TRUE forces exact intervals.

FIGURE 2.4: Empirical CDF and Confidence Bounds for Arsenic in Nails

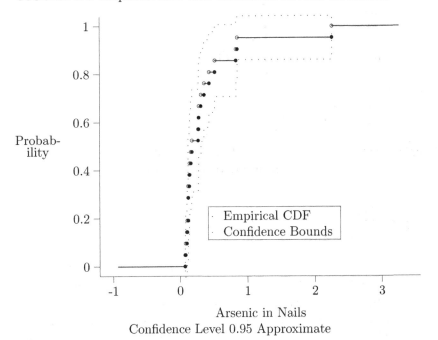

2.6 Exercises

1. Calculate the asymptotic relative efficiency for the sign statistic relative to the one-sample t-test (which you should approximate using the one-sample z-test). Do this for observations from the

    a. uniform distribution, on $[-1/2, 1/2]$ with variance $1/12$ and mean under the null hypothesis of 0, and

    b. logistic distribution, symmetric about 0, with variance $\pi^2/3$ and density $\exp(x)/(1+\exp(x))^2$.

FIGURE 2.5: Empirical CDF and Confidence Bounds for Arsenic in Nails

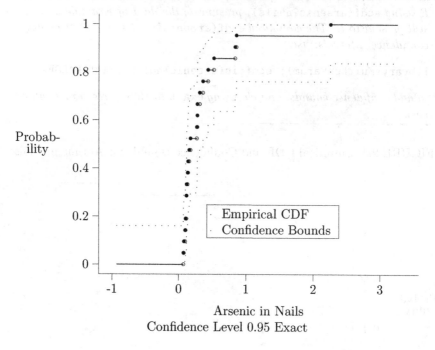

2. The data set

   http://ftp.uni-bayreuth.de/math/statlib/datasets/lupus

   gives data on 87 lupus patients. The third column gives duration, and the fourth column gives transformed disease duration.

   a. Give a 90% confidence interval for the median duration, through inverting the sign test, and compare this to the normal theory interval for the mean. Keep in mind that the normal theory and sign test approach are only comparable if you can argue that the mean and the median for the distribution are plausibly the same. Comment on this.

   b. Give a 90% point-wise confidence bound for the distribution function for disease duration.

3. The data set

   http://lib.stat.cmu.edu/datasets/bodyfat

   gives data on body fat in 252 men. The second column gives proportion of lean body tissue. Give a 95% confidence interval for upper quartile proportion of lean body tissue. Note that the first 116 lines

Exercises 37

and last 10 lines are data set description, and should be deleted. (Line 117 is blank, and should also be deleted).

4. Suppose 49 observations are drawn from a Cauchy distribution, displaced to have location parameter 1.

   a. What is the power of the sign test at level 0.05 to test the null hypothesis of expectation zero for these observations?

   b. What size sample is needed to distinguish between a null hypothesis of median 0 and an alternative hypothesis of median 1, for independent Cauchy variables with a one-sided level 0.025 sign test to give 80% power?

3

Two-Sample Testing

This chapter addresses the question of two-sample testing. Data will generally consist of observations $X_1, \ldots, X_{M_1}$ from continuous distribution function F, and observations $Y_1, \ldots, Y_{M_2}$ from a continuous distribution function G. Model these observations as independent, and unless otherwise specified, treat their distributions as identical, up to some known shift θ; that is,

$$F(z) = G(z - \theta) \; \forall z. \tag{3.1}$$

Techniques for testing a null hypothesis of form $\theta = \theta^0$ in (3.1), vs. the alternative that (3.1) holds for some alternative $\theta \neq \theta^0$, and for estimating θ assuming (3.1), are presented. Techniques for tests in which (3.1) is the null hypothesis, for an unspecified θ, are also presented.

3.1 Two-Sample Approximately Gaussian Inference

Two-sample Gaussian-theory inference primarily concerns expectations; however, one might also compare other aspects of distributions. Under the assumption of an approximate Gaussian distribution, the only additional aspect of the distributions to be compared is their dispersion.

3.1.1 Two-Sample Approximately Gaussian Inference on Expectations

If the observations $X_1, \ldots, X_{M_1}$ and $Y_1, \ldots, Y_{M_2}$ are approximately Gaussian distributed, one might use the test statistic

$$T(\theta) = (\bar{Y} - \bar{X} - \theta)/\sqrt{s_p^2(1/M_2 + 1/M_1)}, \tag{3.2}$$

for

$$S_X^2 = \frac{\sum_{i=1}^{M_1}(X_i - \bar{X})^2}{M_1 - 1}, \; S_Y^2 = \frac{\sum_{i=1}^{M_2}(Y_i - \bar{Y})^2}{M_2 - 1} \tag{3.3}$$

and

$$s_p^2 = \frac{(M_1 - 1)S_X^2 + (M_2 - 1)S_Y^2}{N - 2},$$

for $N = M_1 + M_2$. Statistic (3.2) is called the two-sample pooled t statistic, and the associated test is the two-sample pooled t-test. When θ is correctly specified,

$$T(\theta) \sim \mathfrak{T}_{N-2}, \qquad (3.4)$$

and so the standard test of level α rejects the null hypothesis $\theta = \theta^0$ in (3.1) when

$$|T(\theta^0)| \geq \mathfrak{T}_{N-2,\alpha/2}, \qquad (3.5)$$

for $T(\theta)$ of (3.2). Here $\mathfrak{T}_{N-2,\alpha/2}$ is the $1 - \alpha/2$ quantile of the $\mathfrak{T}_{N-2}$ distribution.

Estimates of θ, under the assumption (3.1), may be constructed by setting $T(\hat{\theta}) = 0$; that is, $\hat{\theta} = \bar{Y} - \bar{X}$. Confidence intervals are generally constructed by inverting the pooled two-sample t-test (3.2) and (3.5) to obtain the interval $\{\theta | |T| \leq \mathfrak{T}_{N-2,\alpha/2}\} = \bar{Y} - \bar{X} \pm s_p \mathfrak{T}_{N-2,\alpha/2} \sqrt{1/M_1 + 1/M_2}$.

3.1.2 Approximately Gaussian Dispersion Inference

One might consider the formerly-alternative hypothesis (3.1), with θ unspecified, as a null hypothesis. Under the Gaussian hypothesis, (3.1) fails to hold only if the variances of the distributions are unequal. In order to compare variances of Gaussian variables, one might compare the separate variance estimates. Under the model (3.1), and with the distributions approximately Gaussian,

$$T = S_Y^2/S_X^2 \sim \mathfrak{f}_{M_2-1, M_1-1} \qquad (3.6)$$

for S_X^2 and S_Y^2 of (3.3) (Fisher, 1925, p. 808), although Fisher (1930) recommended transforming this ratio by taking logs to obtain an approximately Gaussian test statistic. A simpler test, with $\mathrm{E}[X]$ and $\mathrm{E}[Y]$ known, is also available (Fisher, 1926).

Pearson (1931) notes that, under the hypothesis of equal variances, the log of differences in estimated standard deviations is a monotonic transformation of $(1 + M_2 T/M_1)^{-1}$, and that this quantity follows a Pearson I family; he further considers the test arising from comparing this statistic to this exact null sampling distribution. Fisher (1973), crediting Snedecor (1934), recommends comparing T of (3.6) to the $\mathfrak{f}_{M_2-1, M_1-1}$ distribution.

When the underlying distribution of $X_1, \ldots, X_{M_1}$ and $Y_1, \ldots, Y_{M_2}$ is not exactly Gaussian, the distributional result in (3.4) and in (3.6) are approximate rather than exact. This approximation of (3.6) was observed to be poor for even moderate deviations from the Gaussian distribution, in cases when (3.4) remains entirely adequate (Pearson, 1931). Testing for equality of dispersion is revisited in §3.9, and in some sense a nonparametric dispersion test is of more urgency than the test of location.

3.2 General Two-Sample Rank Tests

Nonparametric alternatives to the two-sample pooled t-test, to be developed in this chapter, will reduce to rank tests of the form

$$T_G^{\{k\}} = \sum_{j=1}^{N} a_j I_j^{\{k\}}, \qquad (3.7)$$

for $I_j^{\{k\}}$ equal to the 1 if the item ranked j in the combined sample comes from the group k, and 0 otherwise. The superscript in $I_j^{\{k\}}$ refers to group, and does not represent power.

The statistic $T_G^{\{2\}}$ is designed to take on large values when items in group two are generally larger than the remainder of the observations (that is, the items in group one), and to take small values when items in group two are generally smaller than the remainder of the observations. The statistic $T_G^{\{1\}}$ is designed to take on large values when items in group one are generally larger than the remainder of the observations (that is, the items in group two), and to take small values when items in group one are generally smaller than the remainder of the observations. The statistic $T_G^{\{1\}}$ provides no information not also captured in $T_G^{\{2\}}$, since $T_G^{\{2\}} = \sum_{j=1}^{N} a_j - T_G^{\{1\}}$.

3.2.1 Null Distributions of General Rank Statistics

In order to use this statistic in a hypothesis test, one first needs to know its distribution under the null hypothesis. That is, one needs first to be able to calculate one-sided p-values of form $P_0\left[T_G^{\{2\}} \geq t\right]$, for $T_G^{\{2\}}$ of (3.7) and various values of t, with the subscript 0 on the probability indicating calculation under the null hypothesis. One might calculate p-values for these test statistics exactly: List all $\binom{N}{M_2}$ possible ways to divide the N objects into two groups, one of size M_1. Calculate the test statistic for each rearrangement. Count the number of rearrangements giving a test statistic as extreme or more extreme than the one observed. Divide this count by $\binom{N}{M_2}$ to obtain the p-value. If there exist different rankings giving the same value for $T_G^{\{2\}}$, this information might be used to simplify calculations, but in general one cannot depend on such a simplification. Hence these calculations are likely quite slow. A common simplification is to round the scores so that the scores become integer multiples of a common value; these simplified scores may then be more amenable to exact analysis.

More commonly, the distribution of rank statistics is approximated by the Gaussian distribution. Since $T_G^{\{2\}}$ is the sum of random variables that are neither identically distributed nor independent, a variant of the usual central limit theorem due to Erdös and Réyni (1959), specific to finite population sampling without replacement, is used. There are some conditions on the set of scores needed to ensure that they are approximately Gaussian. Critical values for the two-sided test of level α of the null hypothesis of equality of distribution, vs. the alternative, relying on the Gaussian approximation, are given by

$$\mathrm{E}_0\left[T_G^{\{2\}}\right] \pm \sqrt{\mathrm{Var}_0\left[T_G^{\{2\}}\right]} z_{\alpha/2} \qquad (3.8)$$

and the p-value is given by

$$2\bar{\Phi}\left(|T_G^{\{2\}} - \mathrm{E}_0\left[T_G^{\{2\}}\right]|/\sqrt{\mathrm{Var}_0\left[T_G^{\{2\}}\right]}\right). \qquad (3.9)$$

Moments needed to evaluate (3.8) and (3.9) are given in the next section.

3.2.2 Moments of Rank Statistics

The Gaussian approximation to the null distribution of $T_G^{\{1\}}$ and $T_G^{\{2\}}$ requires calculation of the null expectation and variance of the statistic. This subsection determines moments for a statistic of the form (3.7). Under the null hypothesis, M_1 of these scores are assigned to individuals in the first group, and M_2 scores are assigned to individuals in the the second group, with all rearrangements having equal probability. The scores for the Y group may be thought of as sampled without replacement from a finite population. Let $\bar{a} = \sum_{k=1}^{N} a_k/N$. In this case, $\mathrm{E}\left[T_G^{\{k\}}\right] = \sum_{j=1}^{N} a_j \mathrm{E}\left[I_j^{\{k\}}\right]$, and $\mathrm{E}\left[I_j^{\{k\}}\right] = M_k/N$. Hence

$$\mathrm{E}_0\left[T_G^{\{k\}}\right] = M_k \sum_{j=1}^{N} a_j/N = M_k \bar{a}; \qquad (3.10)$$

the subscript 0 on the expectation operator indicates that the expectation is taken under the null hypothesis.

Calculating the variance is harder, since the $I_j^{\{k\}}$ are not independent. Let $\hat{a} = \sum_{k=1}^{N} a_k^2/N$. Let $b_1 = \mathrm{Var}\left[I_j^{\{k\}}\right] = M_k(N - M_k)/N^2$. When $j \neq i$, then

$$\begin{aligned}
\mathrm{E}\left[I_j^{\{k\}} I_i^{\{k\}}\right] &= \mathrm{P}\left[I_j^{\{k\}} = 1, I_i^{\{k\}} = 1\right] \\
&= \mathrm{P}\left[I_j^{\{k\}} = 1 | I_i^{\{k\}} = 1\right] \mathrm{P}\left[I_i^{\{k\}} = 1\right] = \frac{M_k - 1}{N - 1} \frac{M_k}{N}.
\end{aligned}$$

Let

$$b_2 = \mathrm{Cov}\left[I_2^{\{k\}}, I_1^{\{k\}}\right] = \frac{M_k(M_k - 1)}{N(N - 1)} - \frac{M_k^2}{N^2} = -\frac{(N - M_k)M_k}{N^2(N - 1)}.$$

A First Distribution-Free Test

TABLE 3.1: Reduction of Two-Sample Testing Problem to Fisher's Exact Test, via Mood's Test, for an even total sample size

	Y	X	Total
Greater than Median	A	$N/2 - A$	$N/2$
Less than Median	B	$N/2 - B$	$N/2$
Total	M_2	M_1	N

So

$$\operatorname{Var}\left[T_G^{\{k\}}\right] = \sum_{i=1}^{N}\sum_{j=1}^{N} a_i a_j \operatorname{Cov}\left[I_i^{\{k\}}, I_j^{\{k\}}\right]$$

$$= \sum_{i=1}^{N} a_i^2 b_1 + \sum_{i \neq j} a_i a_j b_2 = (b_1 - b_2)N\hat{a} + N^2 b_2 \bar{a}^2$$

$$= (N - M_k)M_k(\hat{a} - \bar{a}^2)/(N - 1). \tag{3.11}$$

3.3 A First Distribution-Free Test

The nonparametric approach analogous to the sign test is Mood's median test. One first calculates the combined sample median. Let A be number of observations from Y's above the combined median. Let B be number of observations from Y's below the combined median. Under the null hypothesis $F(x) = G(x) \forall x$, and when N is even, A has a hypergeometric distribution, and Mood's test reduces the two-sample equality of distribution problem to Fisher's exact test, as exhibited in Table 3.1. Mood's test T_M is equivalent to the score test $T_G^{\{2\}}$ of (3.7) with

$$a_j = \begin{cases} 1 & \text{for } j \geq (N+1)/2 \\ 0 & \text{for } j = (N+1)/2 \\ -1 & \text{f } \text{ } \text{gj} \leq (N+1)/2, \end{cases} \tag{3.12}$$

although, as originally formulated, this test was applied only in the case of even sample sizes, and so the score 0 would not be used.

Mood's test, and other rank tests, ignore the ordering of the observations from the first group among themselves, and similarly ignore the orderings of the second group among themselves. Represent the data set as a vector of N symbols, M_1 of them X and M_2 of them Y. The letter X in position j indicates that, after ordering a combined set of N observations, the observation ranked j comes from the first group, and the letter Y in position j indicates that the observation ranked j comes from the second group. The advantage of Mood's test lies in its simplicity, and its disadvantage is its low power. To

TABLE 3.2: Reduction of Two-Sample Testing Problem to Fisher's Exact Test, via Mood's Test, for an odd total sample size

	Y	X	Total
Greater than Median	A	$(N-1)/2 - A$	$(N-1)/2$
Equal to Median	C	$1 - C$	1
Less than Median	B	$(N-1)/2 - B$	$(N-1)/2$
Total	M_2	M_1	N

see why its power is low, consider the test with $M_1 = M_2 = 3$, for a total of six observations. A data set label X, Y, X, Y, X, Y indicates that the lowest observation is from the first group, the second lowest is from the second group, the third lowest is from the first group, the fourth lowest is from the second group, the fifth lowest (that is, the second highest) is from the first group, and the highest is from the second group. Mood's test treats X, Y, X, Y, X, Y and X, Y, X, X, Y, Y as having equal evidence against H_0, but the second should be treated as having more evidence. Furthermore, Mood's test takes a value between 0 and $\min(M_2, \lfloor N/2 \rfloor)$. This high degree of discreteness in the statistic's support undermines power.

Westenberg (1948) presented the equal-sample case of the statistic, and Mood (1950) detailed the use in the case with an even combined sample size. Mood's test is of sufficiently low importance in applications that the early references did not bother to present the slight complication that arises when the combined sample size is odd.

When the total sample size is odd, one might represent the median test as inference from a 2 × 3 contingency table with ordered categories, as in Table 3.2. Then $T_M = A - B$. Then the one-sided p-values may be calculated as

$$P_0[A - B \geq t] = P_0[A - B \geq t | C = 0] \frac{M_1}{N} + P_0[A - B \geq t | C = 1] \frac{M_2}{N}.$$

The probabilities $P_0[A - B \geq t | C = c]$ are calculated from the hypergeometric distribution.

The null expectation and variance of T_M are given by (3.10) and (3.11) respectively. Note $\bar{a} = 0$, and

$$\hat{a} = \begin{cases} (N-1)/N & \text{if } N \text{ odd} \\ 1 & \text{if } N \text{ even.} \end{cases}$$

Then (3.10) shows that

$$E_0[T_M] = 0, \quad \text{Var}_0[T_M] = \begin{cases} M_1 M_2 / N & \text{if } N \text{ odd} \\ M_1 M_2 / (N-1) & \text{if } N \text{ even.} \end{cases} \quad (3.13)$$

Critical values and p-values for Mood's test may be calculated from (3.8) and (3.9) respectively.

A First Distribution-Free Test

Example 3.3.1 *Cox and Snell (1981, Example Q) present data on breaking loads (in oz.) of yarn, of two types (A and B) coming from six bobbins. Data may be found at*

http://stat.rutgers.edu/home/kolassa/Data/yarn.dat .

Each combination of bobbin and type is represented four times, for 48 observations in a balanced design. Gaussian quantile plots may be calculated using

```
yarn<-as.data.frame(scan("yarn.dat",what=
   list(strength=0, bobbin=0,type="")))
par(mfrow=c(2,1))
for(yt in c("A","B"))
   qqnorm(yarn$strength[yarn$type==yt],
      pch=yarn$bobbin[yarn$type==yt],
      main=paste("Gaussian QQ plot for Yarn Type",yt))
par(mfrow=c(1,1))
```

and are presented in Figure 3.1. The deviation from a straight line for these points indicates departure from the Gaussian distribution; a potential thresholding associated with bobbin indicates a role for bobbin as a second factor, to be addressed in Chapter 5. Boxplots by type are shown in Figure 3.2. Use Mood's test for a difference in median, ignoring the effect of bobbin. The median strength is 15.75; 15 yarn samples of type B are above the joint median. This leads to Table 3.3, and $T_M = 15 - 9 = 6$. In this even-sample case, from (3.13), $\mathrm{Var}_0[T_M] = 24 \times 24/47 = 12.26$. Hence the approximately standard Gaussian statistic is $(6-0)/\sqrt{12.26} = 1.71$ (and, corrected for continuity, is $(5-0)/\sqrt{12.26} = 1.43$). The correction for continuity here is 1, because possible values of T_M are two units apart. The p-value $2 \times \bar{\Phi}(1.43) = 0.153$. Do not reject the null hypothesis of equal yarn strength. You might also do this using

```
library(MultNonParam)
attach(yarn)
mood.median.test(strength[type=="A"],strength[type=="B"])
```

or

```
genscorestat((strength>median(strength))*2-1,type,correct=1)
```

Compare this with the two-sample t-test:

```
t.test(strength[type=="A"],strength[type=="B"])
detach(yarn)
```

to give a p-value 0.029.

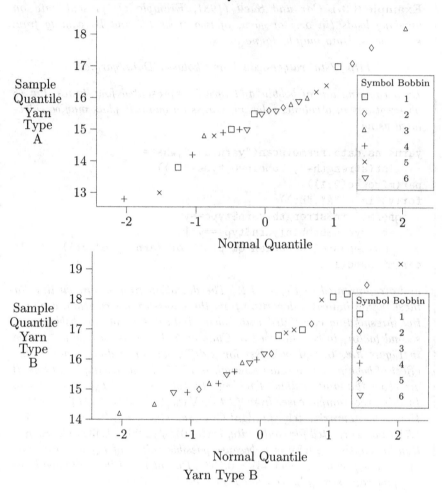

FIGURE 3.1: Normal Quantile Plot for Yarn

Mood's test, using (3.7) in conjunction with (3.12), is almost never used, and is presented here only as a point of departure for more powerful tests. Mood's test looses power primarily because of the discreteness of the scores (3.12). The balance of this chapter explores tests with less discrete scores.

3.4 The Mann-Whitney-Wilcoxon Test

The <u>Wilcoxon rank-sum test</u> is defined as $T_G^{\{2\}}$ of (3.7), with $a_j = j$ (Wilcoxon, 1945). That is, the statistic is the sum of ranks of observations

TABLE 3.3: Mood's test for the yarn example

	Type A	Type B	Total
Greater than Median	9	15	24
Less than Median	15	9	24
Total	24	24	48

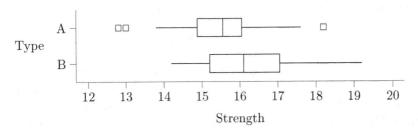

FIGURE 3.2: Boxplots of Yarn Strength (oz) by Type

coming from the second group. Denote this specific example of the general rank statistic as T_W. The term "rank-sum" is used to differentiate this test from another test proposed by the same author, to be discussed in a later chapter.

An alternative test statistic for detecting group differences is

$$T_U = \sum_{i=1}^{M_1} \sum_{j=1}^{M_2} I(X_i < Y_j). \tag{3.14}$$

This new statistic T_U and the previous statistic T_W can be shown to be identical. For $j \in \{1, \ldots, M_2\}$ indexing a member of the second sample, define R_j as the rank of observation j among all of the observations from the combined samples. Note that

$$\begin{aligned} T_W &= \sum_{j=1}^{M_2} R_j = \sum_{j=1}^{M_2} \#(\text{sample entries less than or equal to } Y_j) \\ &= \sum_{j=1}^{M_2} \#(X \text{ values less than or equal to } Y_j) \\ &+ \sum_{j=1}^{M_2} \#(Y \text{ values less than or equal to } Y_j) \end{aligned} \tag{3.15}$$

The first sum in (3.15) is T_U, and the second is $M_2(M_2+1)/2$, and so

$$T_W = T_U + M_2(M_2+1)/2.$$

The test based on T_U is called the Mann-Whitney test (Mann and Whitney, 1947). This statistic is a U statistic; that is, a statistic formed by summing over pairs of observations in a data set.

3.4.1 Exact and Approximate Mann-Whitney Probabilities

The distribution of test statistic (3.14) can be calculated exactly, via recursion (Festinger, 1946). Let $c_W(t, M_1, M_2)$ the number of ways that M_1 symbols X and M_2 symbols Y can be written in a vector to give $T_U = t$, as in §3.3. Then

$$P_{M_1, M_2}[T_U = t] = c_W(t, M_1, M_2) / \binom{N}{M_2}. \tag{3.16}$$

The collection of vectors giving statistic value t can be divided according to whether the last symbol is X or Y. If the last symbol was X, then ignoring this final value, the vector still gives the same statistic value t, and there are $c_W(t, M_1 - 1, M_2)$ such vectors. If the last symbol was Y, then ignoring this final value, the factor gives the statistic value $t - M_1$, and there are $c_W(t - M_1, M_1, M_2 - 1)$ such vectors. So

$$c_W(t, M_1, M_2) = c_W(t, M_1 - 1, M_2) + c_W(t - M_1, M_1, M_2 - 1). \tag{3.17}$$

The recursion stops once either sample size hits zero:

$$c_W(t, M_1, 0) = \begin{cases} 1 & \text{if } t = 0 \\ 0 & \text{if } t \neq 0 \end{cases}, \text{ and } c_W(t, 0, M_2) = \begin{cases} 1 & \text{if } t = 0 \\ 0 & \text{if } t \neq 0 \end{cases}. \tag{3.18}$$

The maximal value for t is

$$N(N+1)/2 - M_1(M_1+1)/2 = M_2(2M_1 + M_2 + 1)/2;$$

hence the recursion can be stopped early by noting that

$$c_W(t, M_1, M_2) = 0 \text{ if } t < M_2(M_2+1)/2 \text{ or if } t > M_2(2M_1+M_2+1)/2. \tag{3.19}$$

A natural way to perform these calculations is with recursive calls to a computer routine to calculate lower-order probabilities, although the algorithm can be implemented without such explicit recursion (Dinneen and Blakesley, 1973).

3.4.1.1 Moments and Approximate Normality

Using this recursion can be slow, and the argument at the end of §3.2.1 can be used to show that the distribution of the test statistic is approximately Gaussian. Fortunately, a central limit theorem applies to this statistic (Erdös and Réyni, 1959).

The Wilcoxon version of the Mann-Whitney-Wilcoxon statistic is given by (3.7) with $a_j = j$. Then $\sum_{j=1}^{N} a_j = \sum_{j=1}^{N} j = N(N+1)/2$, and $\bar{a} = (N+1)/2$. Hence $E_0[T_W]$ is $M_2(N+1)/2$, using (3.10).

The Mann-Whitney-Wilcoxon Test

In order to calculate $\text{Var}_0[T_W]$ using (3.11), one needs $g(w) = \sum_{j=1}^{w} j^2$. One might guess it must be cubic in N. Examine functions $g(w) = aw^3 + bw^2 + cw + d$ so that $g(0) = 0$ and $g(w) - g(w-1) = w^2$. Then $d = 0$, and

$$\begin{aligned} w^2 &= (aw^3 + bw^2 + cw) - \\ &\quad aw^3 + 3aw^2 - 3aw + a - bw^2 + 2bw - b - cw + c \\ &= 3aw^2 - 3aw + a + 2bw - b + c. \end{aligned}$$

Equating quadratic terms above gives $a = 1/3$. Setting the linear term to zero gives $b = 1/2$, and setting the constant term to zero gives $c = 1/6$. Then

$$\sum_{j=1}^{w} j^2 = g(w) = w(2w+1)(w+1)/6, \tag{3.20}$$

$$\hat{a} = (2N+1)(N+1)/6,$$
$$\hat{a} - \bar{a}^2 = (2N+1)(N+1)/6 - (N+1)^2/4 = (N^2-1)/12, \tag{3.21}$$

and from (3.10), (3.11) and (3.21),

$$\text{E}[T_W] = M_2(N+1)/2, \quad \text{Var}[T_W] = M_1 M_2 (N+1)/12. \tag{3.22}$$

In conjunction with the central limit theorem argument described above, one can test for equality of distributions, with critical values and p-values given by (3.8) and (3.9) respectively.

Example 3.4.1 *Refer again to the yarn data of Example 3.3.1. Consider yard strengths for bobbin 3.*

14.2(B), 14.5(B), 14.8(A), 15.2(B), 15.8(A), 15.9(B), 16.0(A), 18.2(A).

Sum the ranks associated with Type B, to get $T_W = 1 + 2 + 4 + 6 = 13$. Here $M_1 = M_2 = 4$, and $N + 1 = M_1 + M_2 + 1 = 9$. From (3.22), under the null hypothesis of equality of distributions, the expected value of the rank sum is $4 \times 9/2 = 18$, and the variance is $4 \times 4 \times 9/12 = 12$. Hence the statistic, after standardizing to zero mean and unit variance, is $(13 - 18)/\sqrt{12} = -1.44$. The $p-$value is 0.149. This may be done using R by

```
wilcox.test(strength~type,data=yarn[yarn$bobbin==3,],
    exact=FALSE, correct=FALSE)
```

The continuity-corrected p-value uses statistic $(13 + 0.5 - 18)/\sqrt{12}$, and is 0.194, and might be done by

```
wilcox.test(strength~type,data=yarn[yarn$bobbin==3,],
    exact=FALSE)
```

Finally, p-values might be calculated exactly using (3.17), (3.18), and (3.19), and in R by

```
wilcox.test(strength~type,data=yarn[yarn$bobbin==3,],
    exact=TRUE)
```

Moments (3.22) apply to the statistic given by scores $a_j = j$. By contrast, the Mann-Whitney statistic T_U is constructed using (3.7) from scores $a_j = j - (N+1)/2$. The variance of this statistic is still given by (3.22); the expectation is $\mathrm{E}\,[T_U] = M_2(N+1)/2 - M_2(M_2+1)/2 = M_2 M_1/2$.

The Wilcoxon variance in (3.22) increases far more quickly than that of Mood's test as the sample size N increases; relative to this variance, the continuity correction is quite small, and is of little importance.

3.4.2 Other Scoring Schemes

One might construct tests using other scores a_j. A variety of techniques are available for use. One could use scores equal to expected value of order statistics from Gaussian distribution; these are called normal scores. Alternatively, one could use scores calculated from the Gaussian quantile function $a_j = \Phi^{-1}(j/(N+1))$ (Waerden, 1952), called van der Waerden scores, or scores of form $a_j = \sum_{i=j}^{N} i^{-1}$ (Savage, 1956), called Savage scores, or scores equal to expected value of order statistics from exponential distribution, called exponential scores. Van der Waerden scores are an approximation to normal scores. Calculating exact probabilities for general score tests, and the difficulties that this entails, was discussed at the end of §3.2.1.

Scores may be chosen to be optimal for certain distributions. Normal scores are optimal for Gaussian observations. Exponential scores are optimal for exponential observations. Original ranks are optimal for logistic observations. Savage scores are optimal for Lehmann alternatives, discussed below at (3.28).

Example 3.4.2 *Consider the nail arsenic data of Example 2.3.2. One might perform an analysis using these scoring methods.*

```
library(exactRankTests)#Gives savage and vw scores
arsenic$savagenails<-as.numeric(cscores(
    arsenic$nails,type="Savage"))
arsenic$vwnails<-as.numeric(cscores(arsenic$nails,
    type="Normal"))
```

The Mann-Whitney-Wilcoxon Test

The Savage scores are

$$0.603, 0.669, 0.850^*, 0.669, -0.056^*, -0.366, 0.902^*,$$
$$0.371, -0.199, 0.794, 0.952, -1.149, -0.816^*, -2.649,$$
$$-1.649, 0.180, -0.566^*, 0.454, 0.069^*, 0.531^*, 0.280^*.$$

Asterisks denote men. The mean of these scores is $\bar{a} = -0.005$, and the mean of the squares is $\hat{a} = 0.823$. Test for equality of arsenic in nails between sexes. Here $M_1 = 8$ and $M_2 = 13$. The expectation and variance of the test statistic are given by (3.10), as $13\bar{a} = -0.065$, and $13 \times 8 \times (\hat{a} - \bar{a}^2)/20 = 4.28$. Sum scores for women, the second gender group; here $k = 2$, and $T_G^{\{k\}} = -1.32$, the sum of scores above without the asterisk. The z-statistic is $(-1.32 - (-0.065))/\sqrt{4.28} = -0.60$. The p-value is 0.548. Do not reject the null hypothesis. These calculations may be done using

```
library(MultNonParam)#Contains genscorestat
genscorestat(arsenic$vwnails,arsenic$sex)
genscorestat(arsenic$savagenails,arsenic$sex)
```

giving the same results for Savage scores, and the p-value 0.7834 for van der Waerden scores.

3.4.3 Using Data as Scores: the Permutation Test

One might instead use the original data as scores. That is, sort the combined data set $(X_1, \ldots, X_{M_1}, Y_1, \ldots, Y_{M_2})$ to obtain $(Z_{(1)}, \ldots, Z_{(N)})$, with $Z_{(i)} \leq Z_{(i+1)}$ for all i; still assuming continuity, each inequality is strict. Then use $a_j = Z_{(j)}$. Hence the test statistic is

$$T_P = \sum_{j=1}^{N} Z_{(j)} I_j^{\{2\}} = \sum_{j=1}^{M_2} Y_j = M_2 \bar{Y}. \tag{3.23}$$

The analysis is performed conditionally on $(Z_{(1)}, \ldots, Z_{(N)})$; note that both the statistic, and its reference distribution, depend on these order statistics.

Compare T_P with the numerator of the two-sample pooled t-test (3.2):

$$\bar{Y} - \bar{X} = \bar{Y} - \frac{N\bar{Z} - M_2\bar{Y}}{M_1} = \frac{N\bar{Y} - N\bar{Z}}{M_1} = N\left(T_P - M_2\bar{Z}\right)/(M_1 M_2),$$

where $\bar{Z} = \sum_{i=1}^{N} Z_{(i)}/N$. The pooled variance estimate for the two-sample t statistic is

$$s_p^2 = \frac{\sum_{j=1}^{M_1}(X_j - \bar{X})^2 + \sum_{j=1}^{M_2}(Y_j - \bar{Y})^2}{N-2}$$

$$= \frac{\sum_{j=1}^{M_1}(X_j - \bar{Z})^2 - M_1(\bar{X} - \bar{Z})^2 + \sum_{j=1}^{M_2}(Y_j - \bar{Z})^2 - M_2(\bar{Y} - \bar{Z})^2}{N-2}$$

$$= \frac{(N-1)s_Z^2 - M_1(\bar{X} - \bar{Z})^2 - M_2(\bar{Y} - \bar{Z})^2}{N-2}.$$

Some algebra shows this to be

$$s_p^2 = s_Z^2 \frac{N-1}{N-2} - \frac{(T_P - M_2\bar{Z})^2(1/M_1 + 1/M_2)}{(N-2)}.$$

Hence, conditional on $(Z_{(1)}, \ldots, Z_{(N)})$, the two-sample pooled t statistic is

$$\frac{\sqrt{(N-2)N}\,(T_P - M_2\bar{Z})}{\sqrt{(s_Z^2(N-1)M_1M_2 - (T_P - M_2\bar{Z})^2 N)}},$$

for s_Z the sample standard deviation of $(Z_{(1)}, \ldots, Z_{(N)})$.

Hence the pooled two-sample t statistic is a strictly increasing function of the score statistic T_P with ordered data used as scores. However, while the pooled t statistic is typically compared to a $\mathcal{T}$ distribution, the rank statistic is compared to the distribution of values arising from random permutations of the group labels; this is the same mechanism that generates the distribution for the rank statistics with scores determined in advance. In the two-sample case, there are $\binom{N}{M_1}$ ways to assign M_1 labels 1, and M_2 labels 2, to the order statistics $(Z_{(1)}, \ldots, Z_{(N)})$. A less-efficient way to think of this process is to specify N labels, the first M_1 of them 1 and the remaining M_2 of them 2, and randomly assign, or randomly permute, $(Z_{(1)}, \ldots, Z_{(N)})$ without replacement; there are $N!$ such assignments, leading to at most $\binom{N}{M_1}$ distinct values. The observed value of T_P is then compared with the sampling distribution arising from this random permutation of values; such a test is called a <u>permutation test</u>. The same permutation concept coincides with the desired reference distribution for all of the rank statistics in this chapter.

Example 3.4.3 *Again consider the nail arsenic data of Example 2.3.2. Recall that there are 21 subjects in this data set, of whom 8 are male. The permutation test testing the null hypothesis of equality of distribution across gender may be performed in R using*

```
library(MultNonParam)
aov.P(dattab=arsenic$nails,treatment=arsenic$sex)
```

to give a two-sided p-value of 0.482. *In this case, all* $\binom{21}{8} = 203490$ *ways to reassign arsenic nail levels to the various groups were considered. The*

TABLE 3.4: Levels for various Two-Sample Two-Sided Tests, Nominal level 0.05, from 100,000 random data sets each, sample size 10 each

Test	Gaussian	Laplace	Cauchy
T-test	0.04815	0.04414	0.01770
Exact Wilcoxon	0.04231	0.04413	0.04424
Approximate Wilcoxon	0.05134	0.05317	0.05318
Normal Scores	0.04693	0.04744	0.04871
Savage Scores	0.04191	0.04340	0.04319
Mood	0.02198	0.02314	0.02314

TABLE 3.5: Powers for various Two-Sample Two-Sided Tests, Nominal level 0.05, from 100,000 random data sets each, sample size 10 each, samples offset by one unit

Test	Gaussian	Laplace	Cauchy
T-test	0.55445	0.35116	0.06368
Approximate Wilcoxon	0.54661	0.41968	0.20765
Normal Scores	0.53442	0.37277	0.16978
Savage Scores	0.47270	0.33016	0.15225

> statistic T_P of (3.23) was calculated for each assignment, this value was subtracted from the null expectation $\bar{Z}$, and the difference was squared to provide a two-sided statistic. The p-value reported is the proportion of these for which the squared differences among the reassignments meets or exceeds that seen in the original data.

3.5 Empirical Levels and Powers of Two-Sample Tests

As in Table 2.1, one might simulate data from a variety of distributions, and compare levels of the various two-sample tests. Results are in Table 3.4.

Table 3.4 shows that the extreme conservativeness of Mood's test justifies its exclusion from practical consideration. We see that the Wilcoxon test, calibrated exactly using its exact null distribution, falls short of the desired level; a less-conservative equal-tailed test would have a level exceeding the nominal target of 0.05. The conservativeness of the Savage Score test is somewhat surprising. The close agreement between the level of the t-test and the nominal level with Gaussian data is as expected, as is the poor agreement between the level of the t-test and the nominal level with Cauchy data.

As was done in Table 2.3, one might perform a similar simulation under the alternative hypotheses to calculate power. In this case, alternative hypotheses were generated by offsetting one group by one unit. Results are in Table 3.5.

Table 3.5 excludes the exact version of the Wilcoxon test and Mood's test, since for these sample sizes ($M_j = 10$ for $j = 1, 2$), they fail to achieve the desired level for any data distribution. The approximate Wilcoxon test has comparable power to that of the t-test under the conditions optimal for the t-test, and also maintains high power throughout.

3.6 Adaptation to the Presence of Tied Observations

The Mann-Whitney-Wilcoxon statistic is designed to be used for variables arising from a continuous distribution. Processes expected to produce data with distinct values, however, sometimes produced tied values, frequently because of limits on measurement precision. Sometimes an observation from the first group is tied with one from the second group. Then the scheme for assigning scores must be modified. Tied observations are frequently assigned scores averaged over the scores that would have been assigned if the data had been distinct; for example, if $Z_{(1)}, \ldots, Z_{(N)}$ are the ordered values from the combination of the two samples, and if $Z_{(j+1)} = Z_{(j)}$, then both observation j and observation $j+1$ are assigned score $(a_j + a_{j+1})/2$. The variance of the test statistic must be adjusted for this change in scores.

When both tied observations come from the first group, or both from the second group, then one might assume that the tie arises because of imprecise measurement of a process that, measured more precisely, would have produced untied individuals. The test statistic is unaffected by assignment of scores to observations according to either of the potential orderings. However, the permutation distribution is affected, because many of the permutations considered will split the tied observations into different groups. Return to variance formula (3.11). The average rank $\bar{a}$ is unchanged by modification of ranks, but the average squared rank $\hat{a}$ changes by $a_j^2 + a_{j+1}^2 - (a_j + a_{j+1})^2/2 = (a_j - a_{j+1})^2/2$. Then, for each pair of ties in the data, the variance (3.11) is reduced by $M_1 M_2 (a_j - a_{j+1})^2/(N-1)$. This process could be continued for triplets, etc., with more complicated expressions for the correction. Lehmann (2006) derives these corrections for generic numbers of replicated values, in the simpler case in which $a_j = j$; in this case, the correction is applied to the simpler variance expression (3.22).

It is simpler, however, to bypass (3.22), and, instead of correcting (3.11), recalculating (3.11) using the new scores.

When the assumption of continuity of the distributions of underlying measurements does not hold, the distribution of rank statistics is no longer independent of the underlying data distribution, since the rank statistic distribution will then depend on the probability of ties. Hence no exact calculation of the form in §3.4.1 is possible.

It was noted at the end of §3.4.1.1 that continuity correction is of little importance in case of rank tests. When average ranks replace the original ranks, the continuity correction argument using Figure 2.2 no longer holds. Potential values of the test statistic are in some cases closer together than 1 unit apart, and, in such cases, the continuity correction might be abandoned.

3.7 Mann-Whitney-Wilcoxon Null Hypotheses

The Mann-Whitney-Wilcoxon test was constructed to test whether the distribution F of the X variables is the same as the distribution G of the Y variables. This null hypothesis implies that $P[X_k \leq Y_j] = 1/2$. Unequal pairs F and G violate the null hypothesis of this test. However, certain distribution pairs violating the null hypothesis fall in the alternative hypothesis, but the Mann-Whitney-Wilcoxon test has no power to distinguish these. This is true if F and G are unequal but symmetric about the same point. In this case, the standard error of the Mann-Whitney-Wilcoxon test statistic (3.11) is no longer correct, and the expectation under this alternative is the same as it is under the null. The same phenomenon arises if $\int_{-\infty}^{\infty} F(y)g(y)\,dy = 1/2$.

As an example, suppose that $Y_j \sim \mathfrak{E}(1)$, $X_i \sim \mathfrak{G}(\theta, 1)$. We now determine the θ for which the above alternative hypothesis has power no larger than the test size. Solve $1/2 = \int_0^\infty (1-\exp(-y))\exp(-(y-\theta)^2/2)(2\pi)^{-1/2}\,dy$ to obtain $\theta = .876$.

3.8 Efficiency and Power of Two-Sample Tests

In this section, consider models of the form (3.1), with the null hypothesis $\theta = \theta^0$. Without loss of generality, one may take $\theta^0 = 0$; otherwise, shift Y_j by θ^0.

Relative efficiency has already been defined for test statistics T_i such that $(T_i - \mu_i(\theta))/(\sigma_i(\theta)/\sqrt{N}) \approx \mathfrak{G}(0,1)$, for N the total sample size. Asymptotic relative efficiency calculations require specification of how M_1 and M_2 move together. Let $M_1 = \lambda N$, $M_2 = (1-\lambda)N$, for $\lambda \in (0,1)$.

3.8.1 Efficacy of the Gaussian-Theory Test

As in the one-sample case, the large sample behavior of this test will be approximated by a version with known variance. Here $\mu(\theta) = \theta$, and $\text{Var}[T] =$

$$\rho^2\left(\frac{1}{M_2}+\frac{1}{M_1}\right)=\rho^2\left(\frac{1}{N(1-\lambda)}+\frac{1}{N\lambda}\right); \text{ hence}$$

$$\sigma(\theta)=\rho\sqrt{1/\lambda+1/(1-\lambda)}=\rho/\zeta,$$

for ρ^2 the variance of each observation, and $\zeta=\sqrt{\lambda(1-\lambda)}$.

For example, suppose that $Y_j \sim \mathfrak{G}(0,1)$, and $X_i \sim \mathfrak{G}(\theta,1)$. In this case, the efficacy is $e=\zeta$.

Alternatively, suppose that the observations are logistically distributed. Each observation has variance is $\pi^2/3$, and the efficacy is $e=\zeta\sqrt{3}/\pi=.551\zeta$.

The analysis of §2.4.1, for tests as in (2.15) and variance scaled as in (2.19), allows for calculation of asymptotic relative efficiency, in terms of the separate efficacies, defined as the ratio of $\mu'(\theta)$ to $\sigma(\theta)$.

3.8.2 Efficacy of the Mann-Whitney-Wilcoxon Test

In order to apply the results for the asymptotic relative efficiency of §2.4, the test statistic must be scaled so that the asymptotic variance is approximately equal to a constant divided by the sample size, and must be such that the derivative of the mean function is available at zero. Using the Mann-Whitney formulation, and rescaling so that the $T=\sum_{i=1}^{M_1}\sum_{j=1}^{M_2} I(X_i<Y_j)/(M_1 M_2)$, then

$$\text{Var}\,[T]=\frac{N+1}{12 M_1 M_2}\approx \frac{1}{N(12\lambda(1-\lambda))},$$

and so

$$\sigma(0)=1/(\sqrt{12}\zeta). \tag{3.24}$$

Also,

$$\mu(\theta)=\mathrm{P}_\theta\,[Y>X]=\mathrm{P}_0\,[Y+\theta>X]=\mathrm{P}_0\,[\theta>X-Y]. \tag{3.25}$$

For example, suppose that $Y_j \sim \mathfrak{G}(0,1)$, and $X_i \sim \mathfrak{G}(\theta,1)$. The differences $X_i-Y_j \sim \mathfrak{G}(\theta,2)$, and so

$$\mu(\theta)=\Phi(\theta/\sqrt{2}). \tag{3.26}$$

Hence $\mu'(0)=1/(2\sqrt{\pi})$. Also, (3.24) still holds, and

$$e=\frac{1}{2\sqrt{\pi}}\sqrt{12}\zeta=\sqrt{3/\pi}\zeta=.977\zeta.$$

Alternatively, suppose that these distributions have a logistic distribution. In this case,

$$\mu(\theta)=\int_{-\infty}^\infty \int_{x-\theta}^\infty \frac{\exp(x)}{(1+\exp(x))^2}\frac{\exp(y)}{(1+\exp(y))^2}\,dy\,dx$$
$$= e^\theta\left(e^\theta-\theta-1\right)\left(e^\theta-1\right)^{-2}, \tag{3.27}$$

and

$$\mu'(0)=1/6,\; e=(1/6)\sqrt{12}\zeta=(1/\sqrt{3})\zeta=.577\zeta.$$

Efficacies for more general rank statistics may be obtained using calculations involving expectations of derivatives of underlying densities, with respect

Efficiency and Power of Two-Sample Tests

TABLE 3.6: Asymptotic relative efficiencies for Two Sample Tests

Test	Pooled T	MWW	ARE of MWW to T
General	$\mu(\theta) = \theta$ $\sigma(\theta) = \varepsilon\zeta^{-1}$ $e = \zeta/\varepsilon$	$\mu(\theta) = P[\theta > X - Y]$ $\sigma(\theta) = \zeta^{-1}/(2\sqrt{3})$ $e = \zeta\mu'(0)(2\sqrt{3})$	$12\varepsilon^2\mu'(0)^2$
Normal, unit variance	$\mu'(0) = 1$ $\sigma(0) = \zeta^{-1}$ $e = \zeta$	$\mu'(0) = (2\sqrt{\pi})^{-1}$ $\sigma(0) = \zeta^{-1}/(2\sqrt{3})$ $e = \sqrt{3/\pi}\zeta$	$\dfrac{3}{\pi} = .95$
Logistic	$\mu'(0) = 1$ $\sigma(0) = \pi\zeta^{-1}/\sqrt{3}$ $e = \zeta\sqrt{3}/\pi$	$\mu'(0) = \frac{1}{6}$ $\sigma(0) = \zeta^{-1}/(2\sqrt{3})$ $e = \zeta/\sqrt{3}$	$\dfrac{\pi^2}{9} = 1.10$

$\zeta = (\lambda(1-\lambda))^{1/2}$, $\varepsilon = \sqrt{\text{Var}[X_i]}$.

to the model parameter, evaluated at order statistics under the null hypothesis, without providing rank expectations away from the null (Dwass, 1956).

3.8.3 Summarizing Asymptotic Relative Efficiency

Table 3.6 contains results of calculations for asymptotic relative efficiencies of the Mann-Whitney-Wilcoxon test to the Pooled t-test. For Gaussian variables, as expected, the Pooled t-test is more efficient, but only by 5%. For a distribution with moderate tails, the logistic, the Mann-Whitney-Wilcoxon test is 10% more efficient.

3.8.4 Power for Mann-Whitney-Wilcoxon Testing

Power may be calculated for Mann-Whitney-Wilcoxon testing, using (2.22) in conjunction with (3.24) for the null variance of the rescaled test, and (3.25), adapted to the particular distribution of interest. Application to Gaussian and Laplace observations are given by (3.26) and (3.27) respectively. Zhong and Kolassa (2017) give second moments for this statistic under the alternative hypothesis, and allow for calculation of $\sigma_1(\theta)$ for non-null θ. The second moment depends not only on the probability (3.25), but also on probabilities involving two independent copies of X and one copy of Y, and of two independent copies of Y and one copy of X. This additional calculation allows the use of (2.17); calculations below involve the simpler formula.

Example 3.8.1 *Consider using two independent sets of 40 observations each to test the null hypothesis of equal distributions vs. the alternative*

> that (3.1) holds, with $\theta = 1$, and with observations having a Laplace distribution. Then, using (3.27), $\mu(1) = e(e-2)(e-1)^{-2} = 0.661$. The function $\mu(\theta)$ has a removable singularity at zero; fortunately the null probability is easily seen to be $1/2$. Then $\lambda = 1/2$, $N = 80$, $\mu(0) = 1/2$, $\mu(1) = 0.661$, $\sigma(0) = 1/\sqrt{12 \times (1/2) \times (1/2)} = 1/\sqrt{3}$. The power for the one-sided level 0.025 test, from (2.22), is $\Phi(\sqrt{80}(0.661 - 0.5)/(1/\sqrt{3}) - 1.96) = \Phi(0.534) = .707$.
>
> One could also determine the total sample size needed to obtain 80% power. Using (2.24), one needs $(1/\sqrt{3})^2 (z_{0.025} + z_{0.2})^2/(0.661 - 0.5)^2 = 151.4$; choose 76 per group.

In contrast with the shift alternative (3.1), one might consider the Lehmann alternative

$$1 - F(z) = (1 - G(z))^k \ \forall z, \qquad (3.28)$$

for some $k \neq 1$. Power calculations for Mann-Whitney-Wilcoxon tests for this alternative have the advantage that power does not depend on the underlying G (Lehmann, 1953).

As noted above, while efficacy calculations are available for more general rank statistics, the non-asymptotic expectation of the test statistic under the alternative is difficult enough that it is omitted here.

3.9 Testing Equality of Dispersion

One can adapt the above rank tests to test whether two populations have equal dispersion, assuming a common center. If one population is more spread out than another, then the members of one sample would tend to lie outside the points from the other sample. This motivates the Siegel-Tukey test. Rank the points, with the minimum getting rank 1, the maximum getting rank 2, then the second to the maximum getting rank 3, the second to the minimum getting rank 4, the third from the minimum getting rank 5 and continuing to alternate. Then sum the ranks associated with one of the samples. Under the null hypothesis, this statistic has the same distribution as the Wilcoxon rank-sum test. Alternately, one might perform the Ansari-Bradley test, by ranking from the outside in, with extremes getting equal rank, and again summing the ranks from one sample.

The Ansari-Bradley test has a disadvantage with respect to the Siegel-Tukey test, in that one can't use off-the-shelf Wilcoxon tail calculations. On the other hand, the Ansari-Bradley test is exactly invariant to reflection.

Testing Equality of Dispersion 59

TABLE 3.7: Yarn data with rankings for testing dispersion

strength	type	ab	st	strength	type	ab	st
12.8	A	1.0	1.00	15.8	A	24.0	47.00
13.0	A	2.0	4.00	15.9	A	22.0	43.67
13.8	A	3.0	5.00	15.9	B	22.0	43.67
14.2	A	4.5	8.50	15.9	B	22.0	43.67
14.2	B	4.5	8.50	16.0	A	19.5	38.50
14.5	B	6.0	12.00	16.0	B	19.5	38.50
14.8	A	7.5	14.50	16.2	A	17.0	33.33
14.8	A	7.5	14.50	16.2	B	17.0	33.33
14.9	A	10.0	19.33	16.2	B	17.0	33.33
14.9	B	10.0	19.33	16.4	A	15.0	30.00
14.9	B	10.0	19.33	16.8	B	14.0	27.00
15.0	A	13.5	26.50	16.9	B	13.0	26.00
15.0	A	13.5	26.50	17.0	A	11.0	21.33
15.0	A	13.5	26.50	17.0	B	11.0	21.33
15.0	B	13.5	26.50	17.0	B	11.0	21.33
15.2	B	16.5	32.50	17.1	A	9.0	18.00
15.2	B	16.5	32.50	17.2	B	8.0	15.00
15.5	A	19.0	37.67	17.6	A	7.0	14.00
15.5	A	19.0	37.67	18.0	B	6.0	11.00
15.5	B	19.0	37.67	18.1	B	5.0	10.00
15.6	A	22.0	43.33	18.2	A	3.5	6.50
15.6	A	22.0	43.33	18.2	B	3.5	6.50
15.6	B	22.0	43.33	18.5	B	2.0	3.00
15.7	A	24.0	48.00	19.2	B	1.0	2.00

Example 3.9.1 *Consider again the yarn data of Example 3.3.1. Test equality of dispersion between the two types of yarn. Ranks are given in Table 3.7, and are calculated in package* NonparametricHeuristic *as*

```
yarn$ab<-pmin(rank(yarn$strength),rank(-yarn$strength))
yarn$st<-round(siegel.tukey.ranks(yarn$strength),2)
yarnranks<-yarn[order(yarn$strength),
   c("strength","type","ab","st")]
```

R functions may be used to perform the test.

```
library(DescTools)#For SiegelTukeyTest
SiegelTukeyTest(strength~type,data=yarn)
yarnsplit<-split(yarn$strength,yarn$type)
ansari.test(yarnsplit[[1]],yarnsplit[[2]])
```

to find the Siegel-Tukey p-value as 0.7179, and the Ansari-Bradley p-value as 0.6786. There is no evidence of inequality of dispersion.

3.10 Two-Sample Estimation and Confidence Intervals

Practitioners often ask what two samples can tell us about how a population location parameter (be it mean, median, or another quantile) differs in two populations. In the restricted case in which the populations are assumed to be the same except for the location parameter, this question does not depend on what location measure is intended.

Our treatment of two-sample confidence intervals will mirror that of the one-sample interval. That is, this parameter will index a family of test statistics, such that the distribution of the family member is independent of the parameter, when the statistic is evaluated at the correct parameter value. Invert the test by determining for which parameters the null hypothesis is not rejected.

Denote the samples as $X_1, \ldots, X_{M_1}$ and $Y_1, \ldots, Y_{M_2}$ as before. Let θ represent the amount by which a location parameter for the population from which the second sample exceeds that of the first sample. Under the assumption (3.1) that the distributions are identical up to shift, then

$$X_1, \ldots, X_{M_1}, Y_1 - \theta, \ldots, Y_{M_2} - \theta \qquad (3.29)$$

all have the same distribution. Then let $T_G^{\{2\}}(\theta)$ be the general rank statistic (3.7) calculated from this data set (3.29).

Most commonly the scores are chosen to make $T_G^{\{2\}}(\theta)$ the Wilcoxon rank-sum statistic, or equivalently the Mann-Whitney statistic, but conceptually this could be done by inverting, for example, Mood's median test or any other rank test. For general scores a_j, $T_G^{\{2\}}(\theta) = \sum_{j=1}^{N} a_j Z_j(\theta)$, where $Z_j(\theta)$ is 1 if item ranked j among (3.29) came from Y, and 0 otherwise.

One can define an estimator as that value of θ that makes this test statistic equal to its null expectation; that is, $\hat{\theta}$ solves

$$T_G^{\{2\}}(\hat{\theta}) = M_2 \bar{a}. \qquad (3.30)$$

Furthermore, one can determine the largest integer t_l and smallest integer t_u such that

$$P_0\left[T_G^{\{2\}}(0) < t_l\right] \leq \alpha/2, \quad P_0\left[T_G^{\{2\}}(0) \geq t_u\right] \leq \alpha/2, \qquad (3.31)$$

in close parallel with definitions of §2.3.2. Then, reject the null hypothesis if $T_G^{\{2\}}(\theta^0) \leq t_L^\circ$ or $T_G^{\{2\}}(\theta^0) \geq t_U^\circ$ for $t_L^\circ = t_l - 1$ and $t_U^\circ = t_u$, and use as the confidence interval

$$\{\theta | t_l \leq T_G^{\{2\}}(\theta^0) < t_u\}. \qquad (3.32)$$

Applying the Gaussian approximation to $T_G^{\{2\}}(\theta)$,

$$t_l, t_u \approx M_2 \bar{a} \pm z_{\alpha/2} \sqrt{\text{Var}\left[T_G^{\{2\}}(0)\right]}. \qquad (3.33)$$

3.10.1 Inversion of the Mann-Whitney-Wilcoxon Test

When a_j are ranks j, $T_G^{\{2\}}(\theta)$ is the Wilcoxon version of the test. The corresponding Mann-Whitney version $T_U(\theta) = \sum_i \sum_j I(X_i < Y_j - \theta)$ gives the estimator and confidence interval end points more easily. In this case, the null expectation of the statistic is $M_1 M_2/2$, and $T_U(\theta) = M_2 M_1/2$ if and only if $M_2 \times M_1$ even, and exactly $M_2 \times M_1/2$ of $V_{ij} = Y_j - X_i$ are greater than θ, or $M_2 \times M_1$ odd, and $(M_2 \times M_1 - 1)/2$ of $V_{ij} = Y_j - X_i$ are greater than θ, $(M_2 \times M_1 - 1)/2$ are less than θ, and one is equal to θ. Hence the estimator (3.30), specific to the Mann-Whitney test, is the median of differences of pairs $Y_j - X_i$. This estimator is given by Hodges and Lehmann (1963), in the same paper giving the analogous estimator for the one-sample symmetric problem of §5.2.1. The confidence interval created by inverting the Mann-Whitney statistic

$$T_U^{\{2\}}(\theta) = \sum_{i=1}^{M_1} \sum_{j=1}^{M_2} I(Y_i - X_j > \theta)$$

is $\{\theta | t_l \le T_U^{\{2\}}(\theta) < t_u\}$, for the largest t_l and smallest t_u as in (3.31), made specific to the Mann-Whitney statistic:

$$\sum_{k=0}^{t_l-1} P_{M_1,M_2}\left[T_U^{\{2\}}(0) = k\right] \le \frac{\alpha}{2}, \quad \sum_{k=t_u}^{M_1 M_2} P_{M_1,M_2}\left[T_U^{\{2\}}(0) = k\right] \le \frac{\alpha}{2}, \quad (3.34)$$

with probabilities given by (3.16). This defines the Hodges-Lehmann estimator in terminology of Higgins (2004) and SAS Institute Inc. (2017). More standard definitions in the one-sample case follows later. One conventionally uses the mean of middle two paired differences for even $M_2 \times M_1$; the theory does not technically require this.

When ties are present, (3.16) no longer gives the exact sampling distribution of $T_U^{\{2\}}(0)$; exact intervals may be determined using the permutation distribution of the tie-adjusted ranks, as in §3.4.3, or (3.33) may be employed.

In this case of confidence intervals derived by inverting the Mann-Whitney (or equivalently Wilcoxon rank-sum) test, using (3.32), θ satisfies $T_U(\theta) = t_l$ if there are t_l pairs $X_i, Y_j - \theta$ such that the second component is greater than the first, and the remainder of the pairs have the inequality reversed, or equivalently if there are t_l pairs with differences $V_{ij} = Y_j - X_i$ such that $V_{ij} > \theta$, and the remainder reversed.

Hence the confidence interval is formed by, first, finding t_l and t_u as in (3.34), and, second, calculating and sorting all $Y - X$ pairs, and, third, reporting differences t_l and t_u on this list. Let $V_{(k)}$ represent ordered value k from the differences V_{ij}. For θ larger than the largest of these differences, $V_{(M_1 M_2)}$, the test statistic $T_U(\theta)$ is zero. Each time θ decreases past one of the pairwise differences, $T_U(\theta)$ increases by one. Hence the confidence interval is $(V_{(M_1 M_2 + 1 - t_u)}, V_{(M_1 M_2 + 1 - t_l)})$. By symmetry, this is $(V_{(t_l)}, V_{(t_u)})$. If (3.33), or some other approximation to the Mann-Whitney critical values, is employed, then the above technique indicates an order statistic that is generally not an integer. In this case, many procedures interpolate between adjacent order statistics.

Note parallels between this construction and that of the confidence interval for the median in §2.3.3. In this section, pairwise differences replace raw observations, and the Wilcoxon quantiles replace those of the binomial distribution. Furthermore, the direction of the test was changed. The rest of the construction is identical.

> **Example 3.10.1** *Again consider the arsenic data of Example 2.3.2. We estimate the median of the distribution of men's nail arsenic levels minus women's arsenic levels, and begin by calculating the $13 \times 8 = 104$ pairwise differences. One woman had an unusually large nail arsenic level; hence there is a cluster of particularly negative differences. Figure 3.3 displays construction of the confidence interval. Confidence intervals given correspond to those order statistics determined from the quantiles of the Mann-Whitney distribution, as in (3.34). These quantiles are $t_l = 25$ and $t_u = 80$. The confidence interval is then entries 25 and 80 among the list of ordered pairwise differences, or $(-0.278, 0.158)$. Figure 3.3 was constructed using*
>
> ```
> library(NonparametricHeuristic)
> invertsigntest(split(arsenic$nails,arsenic$sex))
> ```
>
> *The above analysis did not reflect the fact of a small number of ties among these pairwise differences. The code*
>
> ```
> wilcox.test(nails~sex,data=arsenic,conf.int=TRUE)
> ```
>
> *gives intervals found by using approximation (3.33) to the test critical values, and interpolating between the appropriate order statistics, to obtain an identical result to the same accuracy.*

Compare Figure 3.3 to Figure 2.3. The test statistic in Figure 2.3 is constructed to be non-decreasing in θ; this simplified the discussion of the interval construction, at the cost of employing a slightly non-standard version of the sign test statistic. Figure 3.3 uses a more standard version of the test statistic.

3.11 Tests for Broad Alternatives

The above development motivates different tests for equality of distribution, depending on the kind of departure of greatest interest. For example, a Mann-Whitney test is appropriate for shift alternatives, and an Ansari-Bradley test

Tests for Broad Alternatives 63

FIGURE 3.3: Construction of the Confidence Interval for Median of Differences in Nail Arsenic by Sex

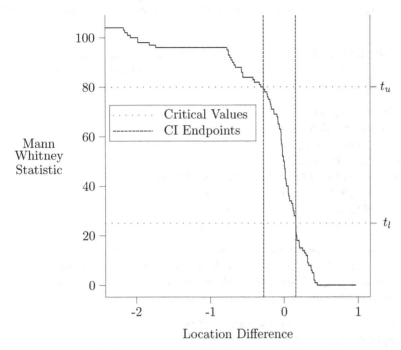

is appropriate for differences in spread. This section describes a test sensitive to departures in various directions.

Use the empirical cumulative distribution function estimator as in §2.5. The Kolmogorov-Smirnov test uses the largest difference between these as the test statistic. The null hypothesis under the permutation distribution is formed from all permutations of data between the two samples. The p-value is the portion with as large or larger difference. If the number of such permutations is quite large, one might use a random sample instead. Asymptotic approximations to these distributions exist as well. Calculation of these statistics can be simplified by noting that the maximum may be calculated from differences in the empirical cumulative distribution functions evaluated exclusively at jumps in one or the other curve.

Alternatively, one might use the integral of difference between these, squared, as the test statistic; this statistic is called the Cramér-von Mises test. That is, if $\hat{F}$ and $\hat{G}$ are the empirical distribution functions for the two samples, then the test statistic is $\int_{-\infty}^{\infty} |\hat{F} - \hat{G}|^2 (M_1 d\hat{F} + M_2 d\hat{G})/N$; this integral is in the sense of Stieltjes (1894), and is calculated as the average over all values in the combined sample of the difference between empirical distributions, squared. Conceptually, to implement this test, use all permutations of data between the two samples, and count the proportion with as large or larger integrated difference as the *p*-value. When the number of permutations is excessive, one might also do this with a random sample of permutations. Alternatively, one might use results from Stochastic Processes to approximate tail areas.

> **Example 3.11.1** *Again consider the the yarn data of Example 3.3.1. Figure 3.4 shows the cumulative distribution functions of strengths for the two types of yarn. This figure might be generated using*
>
> ```
> par(mfrow=c(1,1))
> plot(range(yarn$strength),c(0,1),type="n",
>   main="Yarn Strength",xlab="Yarn Strength",
>   ylab="Probability")
> yarnsplit<-split(yarn$strength,yarn$type)
> lines(ecdf(yarnsplit[[1]]),col=1)
> lines(ecdf(yarnsplit[[2]]),col=2)
> legend(17,.2,lty=c(1,1),col=c(1,2),legend=c("A","B"))
> ```
>
> *The largest difference between distribution functions happens at strengths slightly larger than 16; this gives the Kolmogorov-Smirnov statistic. The Cramér-von Mises statistic is the integral of the squared distance between these curves. Statistics may be calculated using*
>
> ```
> ks.test(yarnsplit[[1]],yarnsplit[[2]])
> library(CvM2SL2Test)
> cvmstat<-cvmts.test(yarnsplit[[1]],yarnsplit[[2]])
> cvmts.pval(cvmstat,length(yarnsplit[[1]]),
>   length(yarnsplit[[2]]))
> ```
>
> *Note that the library* CvM2SL2Test, *giving* cvmts.test *and* cvmts.pval, *is no longer supported, and must be installed from archives, using* library(devtools);install_version("CvM2SL2Test"). *These calculations can also be performed using package* MultNonParam. *Significance tests are based on permutation distributions; the Kolmogorov-Smirnov p-value is 0.2591, and the Cramér-von Mises p-value is 0.09474.*

3.12 Exercises

1. The data set

    http://ftp.uni-bayreuth.de/math/statlib/datasets/schizo

    reflects an an experiment using measurements used to detect schizophrenia, in both schizophrenic and non-schizophrenic patients. Results of various eye-tracking tests are given. The second column

Exercises

FIGURE 3.4: Empirical CDF for Yarn

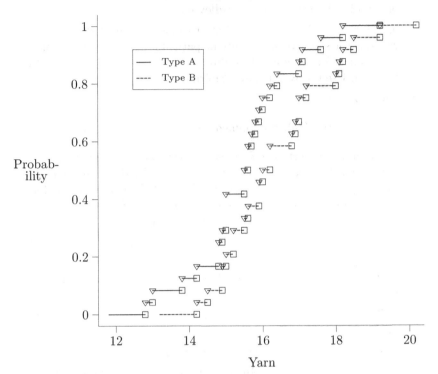

gives eye tracking target type. For all parts of this question select subjects with target type CS. Gain ratios are given in various columns; for the balance of this question, use the first of these, found in the third column.

a. Select only non-schizophrenic patients. The comments at the top of the data file tell you which subjects are female. Compare the first gain ratios for women with those for men. Test whether the gain ratios come from the same distribution, using the Wilcoxon rank sum test.

b. Test whether the first gain ratios for women and men come from the same distribution. Use data values as scores.

For the remainder of this question, select only female patients.

c. Estimate the poplulation median of the difference between the gain ratios for schizophrenic patients (in the lower part of the file) and for the non-schizophrenic patients (in the upper part of the file). Calculate a 95% confidence interval for this median.

d. Test the null hypothesis that the dispersion of the first gain ratios is the same for schizophrenic patients (in the lower part of the file) as it is for the non-schizophrenic patients (in the upper part of the file). Use the Ansari-Bradley test.

e. Test the null hypothesis that the distribution of the first gain ratios is the same for schizophrenic patients (in the lower part of the file) as it is for the non-schizophrenic patients (in the upper part of the file). Use the Kolmogorov-Smirnov test.

2. The data set

$$\text{http://lib.stat.cmu.edu/datasets/biomed.desc}$$

gives values of certain biological markers of a certain disease, on individuals known to be carriers, and control individuals. The carrier subjects are in lines 4 through 78, and the control individuals are in lines 82 through 215. Ignore other lines. Focus attention on column 5 of these data, the first measurement.

a. Graphically assess the quality of a normal model for these measurements.

b. Graphically investigate differences in this variable by carrier/control status.

c. Test the null hypothesis of equality of distribution using the Mann-Whitney-Wilcoxon test.

d. Repeat part (c) using the Normal Scores rank test.

3. Calculate the asymptotic relative efficiency for the Mann-Whitney-Wilcoxon test to the two-sample T-test. Do this for observations from the

a. Laplace distribution of §1.1.1.3.

b. Cauchy distribution of §1.1.1.4, again abusing notation to extend to the Cauchy's property that the central limit theorem does not apply to sums of independent and identically-distributed observations from this distribution.

4. Determine a pair of distributions F and G satisfying both the Lehmann alternative (3.28) and the shift alternative (3.1).

5. Proofs that the two-sample t-statistic for Gaussian data follows the expected distribution uses the fact that the mean differences are independent of the pooled standard deviation estimate, and that the pooled variance estimate times degrees of freedom has a χ^2 distribution with the expected number of degrees of freedom. Using simulation, for tests arising from samples of size $M_1 = M_2 = 10$, evaluate both of these assumptions for data coming from a Laplace distribution, and from a Cauchy distribution.

6. Consider using a one-sided Mann-Whitney-Wilcoxon test of level 0.025 using two samples of size 25, both from a Cauchy distribution.

Exercises

a. Approximate the power for this test to successfully detect an offset of 1 unit.

b. Check the quality of this approximation via simulation.

c. Compare the result from part (a) to the power of the same test, performed two-sided with a level 0.05, to detect the same alternative.

4
Methods for Three or More Groups

This chapter develops nonparametric techniques for one way analysis of variance.

Suppose X_{ki} are samples from K potentially different populations. That is, for fixed k, $X_{k1}, \ldots, X_{kM_k}$ are independent and identically distributed, each with cumulative distribution function F_k. Here $k \in \{1, \ldots, K\}$ indexes group, and M_k represents the number of observations in each group. In order to determine whether all populations are the same, test a null hypothesis

$$H_0 : F_1(x) = \cdots = F_K(x) \forall x, \tag{4.1}$$

vs. the alternative hypothesis H_A : there exists j, k, and x such that $F_j(x) \neq F_k(x)$. Most tests considered in this chapter, however, are most powerful against alternatives of the form $H_A : F_k(x) \leq F_j(x) \forall x$ for some indices k, j, with strict inequality at some k, j, and x. Of particular interest, particularly for power calculations, are alternatives of the form

$$F_i(x - \theta_i) = F_j(x - \theta_j) \tag{4.2}$$

for some constants $\theta_1, \ldots, \theta_K$.

4.1 Gaussian-Theory Methods

Under the assumptions that the data X_{ki} are Gaussian and homoscedastic (that is, having equal variances) under both the null and alternative hypotheses, the null hypothesis H_0 is equivalent to $\mu_j = \mu_k$ for all pairs j, k for $\mu_j = \mathrm{E}[X_{ji}]$. One might test H_0 vs. H_A via analysis of variance (ANOVA). Let $\bar{X}_{k.} = \sum_{i=1}^{M_k} X_{ki}/M_k$, $\bar{X}_{..} = \sum_{k=1}^{K} \sum_{i=1}^{M_k} X_{ki} / \sum_{k=1}^{K} M_k$, and

$$W_A = \frac{(\sum_{k=1}^{K} M_k (\bar{X}_{k.} - \bar{X}_{..})^2)/(K-1)}{\tilde{\sigma}^2}, \tag{4.3}$$

for

$$\tilde{\sigma}^2 = \left(\sum_{k=1}^{K} \sum_{i=1}^{M_k} (X_{ki} - \bar{X}_{k.})^2 \right) / \left(\sum_{k=1}^{K} M_k - K \right). \tag{4.4}$$

When the data have a Gaussian distribution, and (4.1) holds, the numerator and denominator of (4.3) have χ^2 distributions, and are independent; hence the ratio W has an F distribution.

When the data are not Gaussian, the central limit theorem implies that the numerator is still approximately χ^2_{K-1}, as long as the minimal M_k is large, and as long as the distribution of the data is not too far from Gaussian. However, neither the χ^2 distribution for the denominator of (4.3), nor the independence of numerator and denominator, are guaranteed in this case. Fortunately, again for large sample sizes and data not too far from Gaussian, the strong law of large numbers indicates that the denominator of (4.3) is close to the population variance of the observations, and the denominator degree of freedom for the F distribution is large enough to make the F distribution close the the χ^2 distribution. Hence in this large-sample close-to-Gaussian case, the standard analysis of variance results will not mislead.

4.1.1 Contrasts

Let $\mu_k = \mathrm{E}[X_{ki}]$. Continue considering the null hypothesis that $\mu_j = \mu_k$ for all j, k pairs, and consider alternative hypotheses in which X_{ki} all have the same finite variance σ^2, but the means differ, in a more structured way than for standard ANOVA. Consider alternatives such that $\mu_{k+1} - \mu_k$ are the same for all k, and denote the common value by $\Delta > 0$. One might construct a test particularly sensitive to this departure from the null hypothesis using the estimate $\hat{\Delta}$. If $\hat{\Delta}$ is approximately Gaussian, then the associated test of the null hypothesis (4.1) vs. the ordered and equidistant alternative is constructed as $T = (\hat{\Delta} - \mathrm{E}_0[\hat{\Delta}])/\sqrt{\mathrm{Var}_0[\hat{\Delta}]}$; this statistic is compared to the standard Gaussian distribution in the usual way.

An intuitive estimator $\hat{\Delta}$ is the least squares estimators; for example, when $K = 3$ then $\hat{\Delta} = (\bar{X}_{3.} - \bar{X}_{1.})/2$, and when $K = 4$ then $\hat{\Delta} = (3\bar{X}_{4.} + \bar{X}_{3.} - \bar{X}_{2.} - 3\bar{X}_{1.})/10$. Generally, the least squares estimator is a linear combination of group means, of form $\sum_{k=1}^{K} c_k \bar{X}_{k.}$ for a set of constants c_k such that

$$\sum_{k=1}^{K} c_k = 0, \qquad (4.5)$$

with c_k evenly spaced. In this case, $\mathrm{E}_0[\hat{\Delta}] = 0$ and

$$\mathrm{Var}_0[\hat{\Delta}] = \mathrm{Var}_0[X_k] \sum_{k=1}^{K} c_k^2/M_k,$$

and one may use the test statistic

$$T = \sum_{k=1}^{K} c_k \bar{X}_{k.} / \left(\sigma \sqrt{\sum_{k=1}^{K} c_k^2/M_k} \right).$$

Gaussian-Theory Methods

If σ is known, then the null distribution of T is the standard Gaussian distribution. If σ is unknown, an estimator as in (4.4) is substituted, and then the null distribution of T is the $\mathfrak{T}_{N-K}$ distribution.

In this case, W of (4.3) retains its level, but has less power against this ordered alternative.

A linear combination $\sum_{k=1}^{K} c_k \bar{X}_{k\cdot}$ of group means, with constants summing to zero as in (4.5), is called a <u>contrast</u>.

Standard parametric methods will be compared to nonparametric methods below. In order to make these comparisons using methods of efficiency, the pattern of numbers of observations in various groups must be expressed in terms of a single sample size N; presume that $M_k = \lambda_k N$ for all $k \in \{1, \ldots, K\}$. Under alternative (4.2), and with $\varepsilon^2 = \text{Var}[X_{ij}]$ known, then

$$\sum_{k=1}^{K} c_k \bar{X}_k \sim \mathfrak{G}\left(\sum_{k=1}^{K} c_k \theta_k, \varepsilon^2 (\sum_{k=1}^{K} c_k^2/\lambda_k)/N\right).$$

In the case when the shift parameters and the contrast coefficients are equally spaced, and for groups of equal size (that is, $\theta_k = (k-1)\Delta$ for some $\Delta > 0$, $c_k = 2k - (K+1)$, and $\lambda = 1/K$),

$$\sum_{k=1}^{K} c_k \bar{X}_k \sim \mathfrak{G}(K^2(K-1)\Delta/6, (\varepsilon^2/3)K^2(K^2-1)/N).$$

Then $\mu'(\Delta) = (K^2-1)K/6$, and $\sigma(0) = \varepsilon K\sqrt{(K^2-1)/3}$, and the efficacy, as defined in §2.4.1.2, is

$$e = \frac{K(K^2-1)/6}{\varepsilon K\sqrt{(K^2-1)/3}} = \frac{\sqrt{K^2-1}}{2\sqrt{3}\varepsilon}. \tag{4.6}$$

4.1.2 Multiple Comparisons

In the event that the null hypothesis of equal distributions is rejected, one naturally asks which distributions differ. Testing for group differences pairwise (perhaps using the two-sample t-test) allows for $K(K-1)/2$ chances to find a significant result. If each test is done at nominal level, this will inflate the <u>family-wise error rate</u>, or the proportion of experiments that provide any incorrect result. This family-wise error rate will be bounded by the nominal level used for each separate test multiplied by number of possible comparisons performed $(1/2K(K-1))$, but such a procedure, called the <u>Bonferroni procedure</u>, will usually result in a very conservative bound.

Alternatively, consider Fisher's <u>Least Significant Difference</u> (LSD) method. First, perform the standard analysis of variance test (4.3). If this test rejects H_0, then test on each pairwise comparison between mean ranks, and report those pairs whose otherwise uncorrected p-values are smaller than the nominal size. Tests in this second stage may be performed using the two-sample pooled t-test (3.2), except that the standard deviation estimate of (4.4) may be substituted for s_p, with the corresponding increase in degrees of freedom in (3.5).

Fisher's LSD method fails to control family-wise error rate if $K > 3$. To see this, suppose $F_1(x) = F_2(x) = \cdots = F_{K-1}(x) = F_K(x - \Delta)$ for $\Delta \neq 0$. Then null hypotheses $F_j(x) = F_i(x)$ are true, for $i, j < K$. One can make Δ so large that the analysis of variance test rejects equality of all distributions with probability close to 1. Then multiple true null hypotheses are tested, without control for multiplicity. If $K = 3$, there is only one such test with a true hull hypothesis, and so no problem with multiple comparisons.

Contrast this with Tukey's Honest Significant Difference (HSD) method (Tukey, 1953, 1993). Suppose that $Y_j \sim \Phi(0, 1/M_j)$ for $j \in \{1, \ldots, K\}$, $U \sim \chi_m^2$, and that the Y_j and U are independent. Assume further that M_j are all equal. The distribution of $\max_{1 \leq i,j \leq K}(|X_j - X_i|/(\sqrt{U/n}/\sqrt{M_j})$ is called the Studentized range distribution with K and m degrees of freedom. If M_j are not all equal,

$$\sqrt{2} \max_{1 \leq i,j \leq K} ((X_j - X_i)/(\sqrt{U/m}\sqrt{1/M_j + 1/M_k})$$

has the Studentized range distribution with K and m degrees of freedom, approximately (Kramer, 1956); extensions also exist to correlated means (Kramer, 1957). Let $q_{K,m,\alpha}$ be the $1 - \alpha$ quantile of this distribution, and let $\Xi_{K,m}$ be its cumulative distribution function.

One then applies this distribution with $Y_j = (\bar{X}_j - \mu_j)/\sigma$ and U/m the standard sample variance S^2. Here σ is the common standard deviation. If one then sets

$$P_{jk} = \bar{\Xi}_{K,N-K}(\sqrt{2}|\bar{X}_j - \bar{X}_k|/(S\sqrt{(1/M_j + 1/M_k)})), \qquad (4.7)$$

for $N = \sum_{k=1}^{K} M_k$, then for any $\alpha \in (0,1)$,

$$P\left[P_{jk} \leq \alpha \text{ for any } j \neq k \text{ such that } \mu_j = \mu_k\right] \leq \alpha, \qquad (4.8)$$

and the collection of tests that rejects the hypothesis $\mu_i = \mu_j$ if $P_{jk} \leq \alpha$ provides simultaneous test level less than or equal to α. Furthermore, if

$$C_{jk} = \bar{X}_k - \bar{X}_j \pm q_{K,m,\alpha} S\sqrt{1/M_j + 1/M_k}/\sqrt{2}, \qquad (4.9)$$

then

$$P\left[\mu_k - \mu_j \notin C_{jk} \text{ for some } j, k\right] \leq \alpha. \qquad (4.10)$$

This method had been suggested before the Studentized range distribution had been derived (Tukey, 1949).

We now proceed to analogs of one-way analysis of variance that preserve nominal test size for small samples and highly non-Gaussian data.

4.2 General Rank Tests

By analogy with test (4.3), and with the scoring ideas of §3.2, create a test by first ranking all of the observations in a data set, to obtain rank R_{ki} for replicate i in group k. One then creates non-decreasing scores $a_1, \ldots, a_N$, assigns scores $A_{ki} = a_{R_{ki}}$ to the ranked observations, and calculates the score sums $\sum_{i=1}^{M_k} A_{ki}$. One might express the score sums as in (3.7), as $T_G^{\{k\}} = \sum_{j=1}^{N} a_j I_j^{\{k\}}$, for $I_j^{\{k\}}$ equal to the 1 if the item ranked j in the combined sample comes from the group k, and 0 otherwise. Analogously to the numerator in (4.3), let

$$W_G = \sum_{k=1}^{K} u_k \left(T_G^{\{k\}} - \mathrm{E}_0\left[T_G^{\{k\}}\right]\right)^2, \qquad (4.11)$$

for null expectations $\mathrm{E}_0\left[T_G^{\{k\}}\right]$ as calculated in (3.10), and quantities u_k to be determined later. The next subsection will calculate covariances of the $T_G^{\{k\}}$, and the following subsection will demonstrate that W_G has an approximate χ^2_{K-1} null distribution, if

$$u_k = \frac{N-1}{(N^2(\hat{a} - \bar{a}^2))M_k}.$$

The remainder of this section considers the joint distribution of the $T_G^{\{k\}}$, calculates their moments, and confirms the asymptotic distribution for W_G.

4.2.1 Moments of General Rank Sums

First and univariate second moments of $T_G^{\{k\}}$ are as in §3.2.2, and are given by (3.10) and (3.11), respectively. The covariance between $T_G^{\{k\}}$ and $T_G^{\{j\}}$, for $j \neq k$, can be calculated by forming a general rank statistic $T_G^{\{j,k\}}$ combining both groups j and k, to obtain the sum of ranks for individuals in either group j or group k. Note that $T_G^{\{j,k\}} = T_G^{\{j\}} + T_G^{\{k\}}$, and, furthermore, $\mathrm{Var}\left[T_G^{\{j,k\}}\right]$ may be found by applying (3.11), with the number of observations whose ranks are summed being $M_k + M_j$. Then

$$\frac{(N - M_k - M_j)(M_k + M_j)}{(N-1)}(\hat{a} - \bar{a}^2) = \mathrm{Var}\left[T_G^{\{j,k\}}\right]$$
$$= \mathrm{Var}\left[T_G^{\{j\}}\right] + \mathrm{Var}\left[T_G^{\{k\}}\right] + 2\mathrm{Cov}\left[T_G^{\{j\}}, T_G^{\{k\}}\right]$$
$$= \frac{(N-M_j)M_j}{(N-1)}(\hat{a} - \bar{a}^2) + \frac{(N-M_k)M_k}{(N-1)}(\hat{a} - \bar{a}^2) + 2\mathrm{Cov}\left[T_G^{\{j\}}, T_G^{\{k\}}\right]$$

and
$$\operatorname{Cov}\left[T_G^{\{j\}}, T_G^{\{k\}}\right] = \frac{-M_j M_k}{(N-1)}(\hat{a} - \bar{a}^2).$$

4.2.2 Construction of a Chi-Square-Distributed Statistic

This subsection shows that the distribution of W_G is well-approximated by a χ^2_{K-1} distribution.

The use of the Gaussian approximation for the distribution of two-sample general rank statistics was justified at the end of §3.4.1. This argument addressed the distributions of rank sums associated with each group separately; the argument below requires that the joint distribution of the sums of ranks over the various groups be multivariate Gaussian. Hájek (1960), while proving more general distributional finite population sampling results, notes that the results similar to those of Erdös and Réyni (1959) can be applied to all linear combinations of rank sums from separate groups, and hence the collection of rank sums is approximately multivariate Gaussian. This result requires some condition forcing all group proportions to stay away from zero; $\liminf_{N \to \infty} M_k/N > 0$ should suffice.

Statistics formed by squaring observed deviations of data summaries from their expected values, and summing, arise in various contexts in statistics. Often these statistics have distributions that are well-approximated as χ^2. For example, in standard parametric one-way analysis of variance, for Gaussian data, sums of squared centered group means have a χ^2 distribution, after rescaling by the population variance. Tests involving the multinomial distribution also often have the χ^2 distribution as an approximate referent. In both these cases, as with the rank test developed below, the χ^2 distribution has as its degrees of freedom something less than the number of quantities added. In the analysis of variance case, the χ^2 approximation to the test statistic may be demonstrated by considering the full joint distribution of all group means, and integrating out the distribution of the grand mean. In the multinomial case, the χ^2 approximation to the test statistic may be demonstrated by treating the underlying cell counts as Poisson, and conditioning on the table total. In the present case, one might embed the fixed rank sum for all groups taken together into a larger multivariate distribution, and either conditioning or marginalizing, but this approach is unnatural. Instead, below a smaller set of summaries, formed by dropping rank sums for one of the groups, is considered. This yields a distribution with a full-rank variance matrix, and calculations follow.

The total number of observations is $N = \sum_{k=1}^{K} M_k$. Let $\mathbf{Y}$ be the $K-1$ by 1 matrix

$$\left(\frac{T_G^{(1)} - \mathrm{E}\left[T_G^{(1)}\right]}{\sqrt{M_1}}, \ldots, \frac{T_G^{(K-1)} - \mathrm{E}\left[T_G^{(K-1)}\right]}{\sqrt{M_{K-1}}}\right)^\top \omega$$

General Rank Tests

(note excluding the final group rank sum), for $\omega = \sqrt{(N-1)/[N(\hat{a} - \bar{a}^2)]}$. The covariances between components j and k of $\boldsymbol{Y}$ are $-\sqrt{M_j M_k}/N$, and the variance of component j is $1 - M_j/N$. Let $\boldsymbol{\nu}$ be the $K - 1$ by 1 matrix $(\sqrt{M_1/N}, \ldots, \sqrt{M_{K-1}/N})^\top$. Then

$$\text{Var}\,[\boldsymbol{Y}] = \boldsymbol{I} - \boldsymbol{\nu}\boldsymbol{\nu}^\top. \tag{4.12}$$

This proof will proceed by analytically inverting $\text{Var}\,[\boldsymbol{Y}]$. Note that

$$\text{Var}\,[\boldsymbol{Y}]\,(\boldsymbol{I} + (N/M_K)\boldsymbol{\nu}\boldsymbol{\nu}^\top) = \boldsymbol{I} + (-1 + (N/M_K)(1 - \boldsymbol{\nu}^\top\boldsymbol{\nu}))\boldsymbol{\nu}\boldsymbol{\nu}^\top = \boldsymbol{I}, \tag{4.13}$$

since $\boldsymbol{\nu}^\top\boldsymbol{\nu} = \sum_{j=1}^{K-1} M_j/N = 1 - M_K/N$. Then

$$\text{Var}\,[\boldsymbol{Y}]^{-1} = \boldsymbol{I} + (N/M_K)\boldsymbol{\nu}\boldsymbol{\nu}^\top. \tag{4.14}$$

Hence

$$\boldsymbol{Y}^\top(\boldsymbol{I} + (N/M_K)\boldsymbol{\nu}\boldsymbol{\nu}^\top)\boldsymbol{Y} \sim \chi^2_{K-1}. \tag{4.15}$$

Also,

$$\begin{aligned}
\boldsymbol{Y}^\top(\boldsymbol{I} + \frac{N}{M_K}\boldsymbol{\nu}\boldsymbol{\nu}^\top)\boldsymbol{Y} &= \omega^2 \sum_{j=1}^{K-1} (T_G^{\{j\}} - M_j\bar{a})^2/M_j \\
&+ \omega^2 \left(\sum_{j=1}^{K-1}(T_G^{\{j\}} - M_j\bar{a})\right)^2 / M_K \\
&= \omega^2 \left(\sum_{j=1}^{K-1} \frac{(T_G^{\{j\}} - M_j\bar{a})^2}{M_j} + \frac{(T_G^{\{K\}} - M_K\bar{a})^2}{M_K}\right) \\
&= \frac{N-1}{(\hat{a} - \bar{a}^2)N} \sum_{j=1}^{K} \frac{(T_G^{\{j\}} - M_j\bar{a})^2}{M_j}. \tag{4.16}
\end{aligned}$$

The above calculation required some notions from linear algebra. The calculation (4.13) requires an understanding of the definition of matrix multiplication, and the associative and distributive properties of matrices, and (4.14) requires an understanding of the definition of a matrix inverse. Observation (4.15) is deeper; it requires knowing that a symmetric non-negative definite matrix may be decomposed as $\boldsymbol{V} = \boldsymbol{D}^\top \boldsymbol{D}$, for a square matrix $\boldsymbol{D}$, and an understanding that variances matrices in the multivariate case transform as do scalar variances in the one-dimensional case.

One might compare this procedure to either the standard analysis of variance procedure, which is heavily reliant on distribution of responses. Alternatively, one might perform the ANOVA analysis on ranks; this procedure does not depend on distribution of responses.

4.3 The Kruskal-Wallis Test

A simple case of the general multivariate rank statistic (4.16) may be constructed by choosing the scores for the rank statistics to be the identity, with the ranks themselves as the scores.

Kruskal and Wallis (1952) introduced the test that rejects the null hypothesis of equal distributions when the test statistic (4.16) exceeds the appropriate quantile from the null χ^2_{K-1} distribution. They apply this with scores equal to ranks. Using (3.21), $\hat{a} - \bar{a}^2 = (N^2 - 1)/12$, and the statistic simplifies to

$$W_H = (12/[(N+1)N]) \sum_{k=1}^{K} (R_{k.} - M_k(N+1)/2)^2 / M_k. \qquad (4.17)$$

This test is called the Kruskal-Wallis test, and is often referred to as the H test. Here, again, R_{ki} is the rank of X_{ki} within the combined sample, and $R_{k.} = \sum_{i=1}^{M_k} R_{ki}$, and (3.21) gives the first multiplicative factor.

4.3.1 Kruskal-Wallis Approximate Critical Values

Critical values for the Kruskal-Wallis test are often a χ^2_{K-1} quantile. Let $G_k(w; \xi)$ represent the cumulative distribution function for the χ^2 distribution with k degrees of freedom and non-centrality parameter ξ, evaluated at w. Let $G_k^{-1}(\pi, \xi)$ represent the quantile function for this distribution. Then the critical value for the level α test given by statistic (4.17) is

$$G_{K-1}^{-1}(1-\alpha; 0), \qquad (4.18)$$

and the p-value is given by $G_{K-1}(W_H; 0)$.

Example 4.3.1 *The data at*

http://lib.stat.cmu.edu/datasets/Andrews/T58.1

represent corn (maize) yields resulting from various fertilizer treatments (Andrews and Herzberg, 1985, Example 58). Test the null hypothesis that corn weights associated with various fertilizer combinations have the same distribution, vs. the alternative hypothesis that a measure of location varies among these groups. Treatment is a three-digit string representing three fertilizer components. Fields in this file are separated by space. The first three fields are example, table, and observation number. The fourth and following fields are location, block, plot, treatment, ears of corn, and weight of corn. Location TEAN has no ties; restrict attention to that location. The yield for one of the original observations 36 was missing (denoted by -9999 in the file), and is omitted in this analysis. We cal-

The Kruskal-Wallis Test

culate the Kruskal-Wallis test, with 12 groups, and hence 11 degrees of freedom, with 35 observations. Rank sums by treatment are in Table 4.1. Subtracting expected rank sums from observed rank sums, squaring, and dividing by the number of observations in the group gives 971.875. Hence $W_H = 971.875 \times 12/(35 \times 36) = 9.256$. Comparing this to a χ^2_{11} distribution gives the p-value 0.598. Do not reject the null hypothesis of equality of distribution. This might have been done in R using

```
maize<-as.data.frame(scan("T58.1",what=list(exno=0,tabno=0,
    lineno=0,loc="",block="",plot=0,trt="",ears=0, wght=0)))
maize$wght[maize$wght==-9999]<-NA
maize$nitrogen<-as.numeric(substring(maize$trt,1,1))
#Location TEAN has no tied values. R treats ranks of
#missing values nonintuitively. Remove missing values.
tean<-maize[(maize$loc=="TEAN")&(!is.na(maize$wght)),]
cat('\n Kruskal Wallis H Test for Maize Data \n')
kruskal.test(split(tean$wght,tean$trt))
#Alternative R syntax:
#kruskal.test(tean$wght,tean$trt)
```

This might be compared with analysis of variance:

```
#Note that treatment is already a factor.
anova(lm(wght~trt,data=tean))
```

and with analysis of variance of the ranks:

```
anova(lm(rank(wght,na.last=NA)~trt,data=tean))
```

These last two tests have p-values 0.705 and 0.655 respectively. Note the difference between these Gaussian theory results and the Kruskal-Wallis test.

Figure 4.1 shows the support of the normal scores statistic on the set of possible group rank sums for a hypothetical very small data set; the contour of the approximate critical region for the test of level 0.05 is superimposed. As is the case for the chi-square test for contingency tables, points enter the critical region as the level increases in an irregular way, and so constructing an additive continuity correction to (4.17) is difficult. Yarnold (1972) constructs a continuity correction that is additive on the probability, rather on the statistic, scale. Furthermore, even though group sizes are very small, the sample space for the group-wise rank sums is quite rich. This richness of the sample space, as manifest by the small ratio of the point separation (in this case, 1) to the marginal standard deviations (the square roots of the variance in (3.22)), implies that continuity correction will have only very limited utility (Chen and Kolassa, 2018).

TABLE 4.1: Rank sums for the maize production example

Treatment	Replicates	Rank Sum	Expected Rank Sum
000	2	24	36
002	2	48	36
020	1	7	18
022	2	15	36
111	8	167	144
113	4	70	72
131	4	63	72
200	2	38	36
202	2	28	36
220	2	62	36
222	2	42	36
311	4	66	72

4.4 Other Scores for Multi-Sample Rank Based Tests

One might generalize the Kruskal-Wallis test in many of the same ways as one generalized the Mann-Whitney-Wilcoxon test. One might use scoring ideas as before. In (4.17) replace R_{ki} with the scores $a_{R_{ki}}$. Options include van der Waerden scores, Savage scores, and others as described earlier. This provides an adjustment for ties, by letting the scores for the untied entries be the original ranks, and the scores for the tied entries be the average ranks.

Figure 4.2 shows the support of the Kruskal-Wallis statistic on the set of possible normal scores sums for a hypothetical very small data set.

Compare this figure to Figure 4.1, in which sample points for group-wise score sums are far fewer, because more rearrangements of group identifiers lead to the same scores sums. Hence the normal scores distribution shows less discreteness.

Example 4.4.1 *Revisiting the TEAN subset of the maize data of Example 4.3.1, one might perform the van der Waerden and Savage score tests,*

```
library(exactRankTests)#Gives savage, normal scores
library(NonparametricHeuristic)#Gives genmultscore
cat("Other scoring schemes: Normal Scores\n")
genmultscore(tean$wght,tean$trt,
   cscores(tean$wght,type="Normal"))
cat("Other scoring schemes: Savage Scores\n")
genmultscore(tean$wght,tean$trt,
   cscores(tean$wght,type="Savage"))
```

Other Scores for Multi-Sample Rank Based Tests   79

FIGURE 4.1: Asymptotic Critical Region for Kruskal Wallis Test, Level 0.05

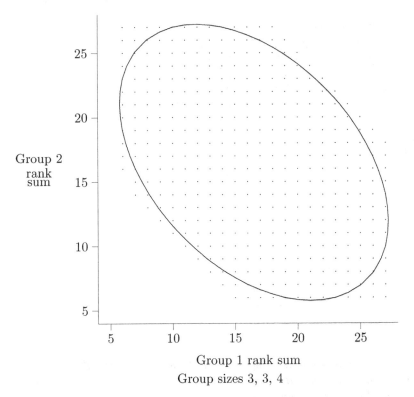

Group 1 rank sum
Group sizes 3, 3, 4

The p-values for normal and Savage scores are 0.9544 and 0.9786 respectively.

One might also apply a permutation test in this context. In this case, the original data are used in place of ranks. The reference distribution arises from the random redistribution of group labels among the observed responses. A Gaussian approximation to this sampling distribution leads to analysis similar to that of an analysis of variance.

Example 4.4.2 *Revisiting the TEAN subset of the maize data of Example 4.3.1, the following syntax performs the exact permutation test, again using package* `MultNonPram`.

```
#date()
#aov.P(tean$wght[!is.na(tean$wght)],
#    tean$trt[!is.na(tean$wght)])
#date()
```

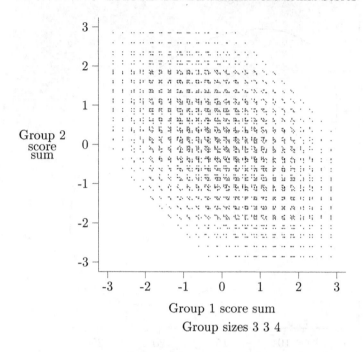

FIGURE 4.2: Distribution of Normal Scores

Group 1 score sum
Group sizes 3 3 4

The date commands bracketing the call to aov.P *allow calculation of elapsed time. However, these calculations are quite slow, and hence are commented out. The commands below approximate the p-value via simulation.*

```
obsp<-anova(lm(wght~trt,data=tean))[[4]][1]
out<-rep(NA,10000)
#Monte Carlo approximation to the permutation distribution
for(i in seq(length(out))){
  out[i]<-anova(lm(sample(wght)~trt,data=tean))[[4]][1]
}
mean(out>=obsp)
```

giving an approximation to the p-value of 0.7254, or, more carefully, $0.7254 \pm 1.96\sqrt{0.7254 \times 0.2746/10000} = (0.717, 0.734)$.

4.5 Multiple Comparisons

The Gaussian-theory multiple comparison techniques of §4.1.2 may be adapted to rank-based testing. The LSD method is adapted by substituting the

Multiple Comparisons 81

Kruskal-Wallis test (4.17) of §4.3 for the analysis of variance test (4.3) in the first stage of the procedure, and substituting the Mann-Whitney-Wilcoxon test (3.14) for the two-sample t-test, with the same lack of Type I error control.

In the case of rank testing, when sample sizes are equal, the Studentized range method may be applied to rank means for simultaneous population differentiation (Dunn, 1964); Conover and Iman (1979) credits this to Nemenyi (1963). This technique may be used to give corrected p-values and corrected simultaneous confidence intervals for rank means. Since rank mean expectations are generally not of interest, the application of the Studentized range distribution to rank means is typically of direct interest only for testing. Use $T_G^{\{k\}}/M_k$ in place of $\bar{X}_k$. Note that for $j \neq k$,

$$\text{Var}\left[\frac{T_G^{\{k\}}}{M_k} - \frac{T_G^{\{j\}}}{M_j}\right] = \frac{\text{Var}\left[T_G^{\{k\}}\right]}{M_k^2} + \frac{\text{Var}\left[T_G^{\{j\}}\right]}{M_j^2} - 2\frac{\text{Cov}\left[T_G^{\{j\}}, T_G^{\{k\}}\right]}{M_j M_k}$$

$$= \left(\frac{N-M_j}{M_j} + \frac{N-M_k}{M_k} + 2\right)\frac{\hat{a} - \bar{a}^2}{N-1}$$

$$= \left(\frac{1}{M_j} + \frac{1}{M_k}\right)\frac{N(\hat{a} - \bar{a}^2)}{N-1}.$$

Hence S in (4.7) may be replaced with $\sqrt{N(\hat{a} - \bar{a}^2)/(N-1)}$ to obtain simultaneous p-values to satisfy (4.8). Also, take the denominator degrees of freedom to be ∞ as the second argument to q. The same substitution may be made in (4.9) to obtain (4.10), but the parameters bounded in these intervals are differences in average rank, which are seldom of interest.

Example 4.5.1 *Consider again the yarn data of Example 3.3.1. Consider just type A, and explore pairwise bobbin differences. One might do all pairwise Mann-Whitney-Wilcoxon tests.*

```
yarna<-yarn[yarn$type=="A",]
cat('\nMultiple Comparisons for Yarn with No Correction\n')
pairwise.wilcox.test(yarna$strength,yarna$bobbin,exact=F,
   p.adjust.method="none")
```

This gives pairwise p-values

```
       1     2     3     4     5
2  0.112   -     -     -     -
3  0.470 1.000   -     -     -
4  0.245 0.030 0.112   -     -
5  0.885 0.312 0.772 0.470   -
6  0.661 0.146 0.470 0.042 1.000
```

Bobbins 2 and 4 seem to be the most different, followed by 4 and 6, but the above comparison is not adjusted for multiple comparisons.

One might perform the version of the Fisher LSD approach using the Kruskal-Wallis test for pairwise comparisons, as described by Higgins (2004):

```
library(MultNonParam)
higgins.fisher.kruskal.test(yarna$strength,yarna$bobbin)
```

In this case, the initial Kruskal-Wallis test fails to reject the null hypothesis of equality of distribution, and no further exploration is performed.

Applying the Bonferroni adjustment,

```
# pairwise.wilcox.test corrects for multiple comparisons
# methods using only on p-values.  This requirement
# excludes methods of Tukey and Scheffe.
cat('\nBonferroni Comparisons for Yarn Type A Data\n')
pairwise.wilcox.test(yarna$strength,yarna$bobbin,
   exact=F,p.ajust.method="bonferroni")
```

This gives corrected p-values:

```
    1    2    3    4    5
2 1.00 -    -    -    -
3 1.00 1.00 -    -    -
4 1.00 0.46 1.00 -    -
5 1.00 1.00 1.00 1.00 -
6 1.00 1.00 1.00 0.59 1.00
```

None are significant after correction for multiplicity. One might also use Tukey's method for calculating p-values respecting multiple comparisons:

```
library(MultNonParam)
tukey.kruskal.test(yarna$strength,yarna$bobbin)
```

indicating no significant differences.

4.6 Ordered Alternatives

Again consider the null hypothesis of equal distributions. Test this hypothesis vs. the ordered alternative hypothesis

$$H_A : F_i(x) \geq F_{i+1}(x) \forall x$$

for all indices $i \in \{1, \ldots, K-1\}$, with strict equality at some x and some i. This alternative reduces parameter space to $1/2^K$ of former size. One might

Ordered Alternatives

use as the test statistic

$$J = \sum_{i<j} U_{ij}, \qquad (4.19)$$

where U_{ij} is the Mann-Whitney-Wilcoxon statistic for testing groups i vs. j.

Reject the null hypothesis when J is large. This statistic may be expressed as $\sum_{k=1}^{K} c_k \bar{R}_k$, plus a constant, for some c_k satisfying (4.5); that is, J may be defined as a contrast of the rank means, and the approach of this subsection may be viewed as the analog of the parametric approach of §4.1.1.

Critical values for J can be calibrated using a Gaussian approximation. Under the null hypothesis, the expectation of J is

$$E_0[J] = \sum_{i<j} M_i M_j / 2 = N^2/4 - \sum_i M_i^2/4,$$

and the variance is

$$\text{Var}_0[J] = \frac{1}{12} \sum_{i=2}^{K} \text{Var}_0[U_i] = \frac{1}{12} \sum_{i=2}^{K} M_i m_{i-1}(m_i + 1); \qquad (4.20)$$

here U_i is the Mann-Whitney statistic for testing group i vs. all preceding groups combined, and $m_i = \sum_{j=1}^{i} M_j$. The second equality in (4.20) follows from independence of the values U_i (Terpstra, 1952). A simpler expression for this variance is

$$\text{Var}_0[J] = \frac{1}{72} \left[N(N+1)(2N+1) - \sum_{i=1}^{K} M_i(M_i+1)(2M_i+1) \right]. \qquad (4.21)$$

This test might be corrected for ties, and has certain other desirable properties (Terpstra, 1952).

Jonckheere (1954), apparently independently, suggested a statistic that is twice J, centered to have zero expectation, and calculated the variance, skewness, and kurtosis. The resulting test is generally called the Jonckheere-Terpstra test.

Example 4.6.1 *Consider again the Maize data from area TEAN in Example 4.3.1. The treatment variable contains three digits; the first indicated nitrogen level, with four levels, and is extracted in the code in Example 4.3.1. Apply the Jonckheere-Terpstra test:*

```
library(clinfun)# For the Jonckheere-Terpstra test
jonckheere.test(tean$wght,tean$nitrogen)
cat('\n K-W Test for Maize, to compare with JT  \n')
kruskal.test(tean$wght,tean$nitrogen)
```

> to perform this test, and the comparative three degree of freedom Kruskal-Wallis test. The Jonckheere-Terpstra Test gives a p-value 0.3274, as compared with the Kruskal-Wallis p-value 0.4994.

4.7 Powers of Tests

This section considers powers of tests calculated from linear and quadratic combinations of of indicators

$$I_{ik,jl} = \begin{cases} 1 & \text{i } X_{ik} < X_{jl} \\ 0 & \text{i } X_{ik} > X_{jl} \end{cases}.$$

The Jonckheere-Terpstra statistic (4.19) is of this form, as is the Kruskal-Wallis statistic (4.17), since the constituent rank sums can be written in terms of pairwise variable comparisons. Powers will be expressed in terms of the expectations

$$\kappa_{ij} = P_A[X_{i1} < X_{j1}] \text{ for } i \neq j, \text{ and } \kappa_{ii} = 1/2. \tag{4.22}$$

Under the null hypothesis of equal populations, $\kappa_{ij} = 1/2$ for all $i \neq j$.

In the case of multidimensional alternative hypotheses, effect size and efficiency calculations are more difficult than in earlier one-dimensional cases. In the case with K ordered categories, there are effectively $K - 1$ identifiable parameters, since, because the location of the underlying common null distribution for the data is unspecified, group location parameters θ_j can all be increased or decreased by the same constant amount while leaving the underlying model unchanged. On the other hand, the notion of relative efficiency requires calculating an alternative parameter value corresponding to, at least approximately, the desired power, and as specified by (2.23). This single equation can determine only a single parameter value, and so relative efficiency calculations in this section will consider alternative hypotheses of the form

$$\boldsymbol{\theta}^A = \Delta \boldsymbol{\theta}^\dagger \tag{4.23}$$

for a fixed direction $\boldsymbol{\theta}^\dagger$. Arguments requiring solving for the alternative will reduce to solving for Δ. The null hypothesis is still $\boldsymbol{\theta} = \mathbf{0}$.

4.7.1 Power of Tests for Ordered Alternatives

The statistic J of (4.19) has an approximate Gaussian distribution, and so powers of tests of ordered alternatives based on J are approximated by (2.18). Under both null and alternative hypotheses,

$$E[J] = \sum_{i<j} M_i M_j \kappa_{ij}, \tag{4.24}$$

Powers of Tests 85

with κ_{ij} defined as in (4.22). Alternative values for κ_{ij} under shift models (4.2) are calculated as in (3.25). Without loss of generality, one may take $\theta_1 = 0$.

Consider parallels with the two-group setup of §3. The cumulative distribution function F_1 of (4.2) corresponds to F of (3.1), and F_2 corresponds to G of (3.1). Then $\mu(\theta)$ of (3.25) corresponds to κ_{12}. Calculation of κ_{kl}, defined in (4.22), and applied to particular pairs of distributions, such as the Gaussian in (3.26) and the logistic in (3.27), and other calculations from the exercises of §3, hold in this case as well. Each of the difference probabilities κ_{kl}, for $k \neq l$, depends on the alternative distribution only through $\theta_l - \theta_k$.

Power may be calculated from (2.18).

Example 4.7.1 *Consider $K = 3$ groups of observations, Gaussian, with unit variance and expectations $\theta_1 = 0$, $\theta_2 = 1/2$, and $\theta_3 = 1$, and all groups of size $M_i = 20$. Consider a one-sided level 0.025 test. Applying (3.26), $\kappa_{12} = \kappa_{23} = \Phi(.5/\sqrt{2}) = 0.638$, and $\kappa_{13} = \Phi(1/\sqrt{2}) = 0.760$. The null and alternative expectations of J are*

$$60^2[(1/3)(1/3) \times 0.5 + (1/3)(1/3) \times 0.5 + (1/3)(1/3) \times 0.5] = 600$$

and

$$60^2[(1/3)(1/3) \times 0.638 + (1/3)(1/3) \times 0.638 + (1/3)(1/3) \times 0.760] = 814.6$$

respectively, from (4.24). The null variance of J, from (4.21), is

$$\begin{aligned}\varsigma^2(0) &= \frac{1}{72}\left[N(N+1)(2N+1) - \sum_{i=1}^{K} M_i(M_i+1)(2M_i+1)\right] \\ &= (60 \times 61 \times 1211 - 3 \times 20 \times 21 \times 41)/72 = 5433.3.\end{aligned}$$

Applying (2.18), power is $1 - \Phi((600 - 814.6)/\sqrt{5433.3} + 1.96) = 0.829$. I used (2.18), rather than (2.20), since the variance of the distribution of the statistic was most naturally given above without division by sample size, and rather than (2.17), because calculating the statistic variance under the alternative is tedious.

This may be computed in R using

```
library(MultNonParam)
terpstrapower(rep(20,3),(0:2)/2,"normal")
```

and the approximate power may compared to a value determined by Monte Carlo; this value is 0.857.

4.7.2 Power of Tests for Unordered Alternatives

Power for unordered alternatives is not most directly calculated as an extension of (2.22). In this unordered case, as noted above, the approximate null distribution for W_H is χ^2_{K-1}. One might attempt to act by analogy with (2.17), and calculate power using alternative hypothesis expectation and variance matrix. This is possible, but difficult, since W_H (and the other approximately χ^2_{K-1} statistics in this chapter) are standardized using the null hypothesis variance structure. Rescaling this under the alternative hypothesis gives a statistic that may be represented approximately as the sum of squares of independent Gaussian random variables; however, not only do these variables have non-zero means, which is easily addressed using a non-central χ^2 argument as in (1.3), but also have unequal variances for the variables comprising the summands; an analog to (1.3) would have unequal weights attached to the squared summands.

A much easier path to power involves analogy with (2.22): Approximate the alternative distribution using the null variance structure. This results in the non-central chi-square approximation using (1.4).

Because the Mann-Whitney and Wilcoxon statistics differ only by an additive constant, the Kruskal-Wallis test may be re-expressed as

$$\left\{\sum_{k=1}^{K}(T_k - M_k(N-M_k)\kappa^\circ)^2/M_k\right\}/[\psi^2(N+1)N]. \qquad (4.25)$$

Here $\kappa^\circ = 1/2$; this is the null value of κ_{kl}, and the null hypothesis specifies that this does not depend on k or l, and $\psi = 1/\sqrt{12}$, a multiplicative constant arising in the variance of the Mann-Whitney-Wilcoxon statistic. Many of the following equations follow from (4.25); furthermore, (4.25) also approximately describes other statistics to be considered later, and analogous consequences may be drawn for these statistics as well, with a different value for ψ. Hence the additional complication of leaving a variable in (4.25) whose value is known will be justified by using consequences of (4.25) later without deriving them again. Here,

$$T_k = \sum_{j=1}^{M_k}\sum_{l=1,l\neq k}^{K}\sum_{i=1}^{M_l} I(X_{kj} > X_{li}), \qquad (4.26)$$

the Mann-Whitney statistic for testing whether group k differs from all of the other groups, with all of the other groups collapsed.

The variance matrix for rank sums comprising W_H is singular (that is, it does not have an inverse), and the argument justifying (1.4) relied on the presence of an inverse. The argument of §4.2.2 calculated the appropriate quadratic form, dropping one of the categories to obtain an invertible variance matrix, and then showed that this quadratic form is the same as that generating t. The same argument shows that the appropriate non-centrality parameter is

$$\xi = \left\{\sum_{k=1}^{K}(\mathrm{E}_A[T_k] - M_k(N-M_k)(1/2))^2/M_k\right\}/[\psi^2(N+1)N],$$

Powers of Tests

where $E_A[T_k] = M_k \sum_{l=1, l \neq k}^{K} M_l \kappa_{kl}$. The non-centrality parameter is

$$\xi = \frac{1}{\psi^2(N+1)N} \sum_{k=1}^{K} M_k \left(\sum_{l=1, l \neq k}^{K} M_l(\kappa_{kl} - \kappa^\circ) \right)^2$$

$$= \frac{1}{\psi^2(N+1)N} \sum_{k=1}^{K} M_k \left(\sum_{l=1}^{K} M_l(\kappa_{kl} - \kappa^\circ) \right)^2. \quad (4.27)$$

The restriction on the range of the inner summation $l \neq k$ may be dropped in (4.27), because the additional term is zero.

Let G_k and G_k^{-1} be the chi-square cumulative distribution and quantile functions respectively, as in §4.3.1. The power for a the Kruskal-Wallis test with K groups, under alternative (4.2), is approximately

$$1 - G_{K-1}(G_{K-1}^{-1}(1 - \alpha, 0); \xi), \quad (4.28)$$

with ξ given by (4.27).

Example 4.7.2 *Continue example 4.7.1. Again, consider $K = 3$ groups of observations, Gaussian, with unit variance and expectations $\theta_1 = 0$, $\theta_2 = 1/2$, and $\theta_3 = 1$, and all groups of size $M_i = 20$. The three inner sums in (4.27) are $20 \times (0.5 - 0.5) + 20 \times (0.638 - 0.5) + 20 \times (0.760 - 0.5) = 7.968$, $20 \times (0.362 - 0.5) + 20 \times (0.5 - 0.5) + 20 \times (0.638 - 0.5) = 0$, and $20 \times (0.240 - 0.5) + 20 \times (0.362 - 0.5) + 20 \times (0.5 - 0.5) = -7.968$. Squaring, multiplying each of these by $M_k = 20$, and adding gives 2539.6. Multiplying by $12/(60 \times 61)$ gives 8.327. The critical value for a test of level 0.05 is given by the χ^2 distribution with 2 degrees of freedom, and is 5.99. The tail probability associated with the non-central χ^2 distribution with non-centrality parameter 8.327 and two degrees of freedom, beyond 5.99, is 0.736; this is the power for the test. As expected, this power is less than that given in example (4.7.1) for the Jonckheere-Terpstra test. This might be compared with a Monte Carlo approximation of 0.770.*

This may be computed using the R package `MultNonParam` *using*

```
kwpower(rep(20,3),(0:2)/2,"normal")
```

Approximation (4.27) may be approximated to give a simpler relation between the non-centrality parameter and sample size, allowing for the calculation of the sample size producing a desired power, denoted in this subsection as $1 - \beta$. In (4.28), sample size enters only through the non-centrality parameter. As in standard one-dimensional sample size calculations, re-express the relation between power and non-centrality as

$$G_{K-1}^{-1}(\beta; \xi) = G_{K-1}^{-1}(1 - \alpha, 0). \quad (4.29)$$

From (4.27),
$$\xi \approx \left\{ \sum_{k=1}^{K} \lambda_k \left(\sum_{l=1}^{K} \lambda_l (\kappa_{kl} - \kappa^\circ) \right)^2 \right\} N/\psi^2, \qquad (4.30)$$
for $\psi = 1/\sqrt{12}$, and
$$N \approx \xi \psi^2 / \left\{ \sum_{k=1}^{K} \lambda_k \left(\sum_{l=1}^{K} \lambda_l (\kappa_{kl} - \kappa^\circ) \right)^2 \right\}, \qquad (4.31)$$
for $\lambda_k = \lim_{N \to \infty} M_k/N$. Assume that $\lambda_k > 0$ for all k. Hence, to determine the sample size needed for a level α test to obtain power $1-\beta$, for an alternative with group differences κ_{kl}, and with K groups in proportions λ_k, first solve (4.29) for ξ, and then apply (4.31). An old-style approach to solving (4.29) involves examining tables of the sort in Haynam et al. (1982a) and Haynam et al. (1982b).

> **Example 4.7.3** *Again, consider $K = 3$ groups of observations, Gaussian, with unit variance and expectations $\theta_1 = 0$, $\theta_2 = 1/2$, and $\theta_3 = 1$ Calculate the sample size needed for the level α Kruskal-Wallis test involving equal-sized groups to obtain power 0.8. Quantities κ_{kl} were calculated in example 4.7.1 to be 0.362, 0.5, and 0.638. The three inner sums in (4.31) are $(0.5-0.5)/3 + (0.638-0.5)/3 + (0.760-0.5)/3 = 0.133$, $(0.362-0.5)/3 + (0.5-0.5)/3 + (0.638-0.5)/3 = 0$, and $(0.240-0.5)/3 + (0.362-0.5)/3 + (0.5-0.5)/3 = -0.133$. Squaring, multiplying each of these by $\lambda_k = 1/3$, adding, and multiplying by 12 gives 0.1408. Solving the equation (4.29) gives $\xi = 9.63$, and the sample size is $9.63/0.1408 \approx 69$, indicating 23 subjects per group.*
>
> *This may be computed using the R package* `MultNonParam` *using*
>
> ```
> kwsamplesize((0:2)/2,"normal")
> ```

As in the one-dimensional case, one can solve for effect size by approximating the probabilities as linear in the shift parameters. Express
$$\kappa_{kl} - \kappa^\circ \approx \kappa'(\theta_k^A - \theta_l^A), \qquad (4.32)$$
and explore the multiplier Δ from (4.23) giving the alternative hypothesis in a given direction. Then
$$N \approx \xi \psi^2 / \left\{ (\kappa')^2 \sum_{k=1}^{K} \sum_{j=1}^{K} \sum_{l=1}^{K} \lambda_k \lambda_j \lambda_l (\theta_k^A - \theta_l^A)^2 \right\} = \frac{\xi \psi^2}{\Delta^2 (\kappa')^2 \zeta^2}, \qquad (4.33)$$
for $\zeta = \sqrt{\left[\sum_{k=1}^{K} \lambda_k \left(\theta_k^\dagger\right)^2 - \left(\sum_{k=1}^{K} \lambda_k \theta_k^\dagger\right)^2 \right]}$; ζ plays a role analogous to its role in §3.8, except that here it incorporates the vector giving the direction of departure from the null hypothesis.

Powers of Tests

Example 4.7.4 *The sum with respect to j disappears from ζ^2, since the sum of the proportions λ_j is 1. Under the same conditions in Example 4.7.3, one might use (4.33) rather than (4.31). Take $\boldsymbol{\theta}^\dagger = \boldsymbol{\theta}^A = (0, 1/2, 1)$ and $\Delta = 1$. The derivative κ' is tabulated, for the Gaussian and logistic distributions, in Table 3.6 as $\mu'(0)$. In this Gaussian case, $\kappa' = \mu'(0) = (2\sqrt{\pi})^{-1} = 0.282$. Also, $\zeta^2 = (0^2/3 + (1/2)^2/3 + 1^2/3 - (0/3 + (1/2)/3 + 1/3)^2) = 5/12 - 1/4 = 1/6$. The non-centrality parameter, solving (4.29), is $\xi = 9.63$. The approximate sample size is $9.63/(12 \times 1^2 \times 0.282^2/6) = 61$, or approximately 21 observations per group; compare this with an approximation of 23 per group using a Gaussian approximation, without expanding the κ_{ij}.*

This may be computed using the R package `MultNonParam` *using*

```
kwsamplesize((0:2)/2,"normal",taylor=TRUE)
```

Relation (4.33) may be solved for the effect size Δ, to yield

$$\Delta = \psi\sqrt{\xi}/\left(\kappa'\zeta N^{1/2}\right). \qquad (4.34)$$

In order to determine the effect size necessary to give a level α test power $1 - \beta$ to detect an alternative hypothesis $\Delta\boldsymbol{\theta}^\dagger$, determine ξ from (4.28), and then Δ from (4.34).

Example 4.7.5 *Again use the same conditions as in Example 4.7.3. Consider three groups with 20 observations each; then $N = 60$. Recall that $\psi = 1/\sqrt{12}$. Example 4.7.4 demonstrates that $\xi = 9.63$, $\kappa' = 0.282$, $\zeta^2 = 1/6$, and $\Delta = (\sqrt{9.63/(60 \times 12 \times (1/6))})/0.282 = 1.004$. As Δ is almost exactly 1, the alternative parameter vector in the direction of $(0, 1/2, 1)$ and corresponding to a level 0.05 test of power 0.80 with 60 total observations is $(0, 1/2, 1)$.*

This may be computed using the R package `MultNonParam` *using*

```
kweffectsize(60,(0:2)/2,"normal")
```

Figure 4.3 reflects the accuracy of the various approximations in this section. It reflects performance of approximations to the power of the Kruskal-Wallis test. Tests with $K = 3$ groups, with the same number of observations per group, with tests of level 0.05, were considered. Group sizes between 5 and 30 were considered (corresponding to total sample sizes between 15 and 90). For each sample size, (4.34) was used to generate an alternative hypothesis with power approximately 0.80, in the direction of equally-spaced alternatives. The dashed line represents a Monte Carlo approximation to the power, based on 50,000 observations. The dotted line represents the standard non-central chi-square approximation (4.28). The solid line represents this same approximation, except also incorporating the linear approximation (4.32) representing the exceedance probabilities as linear in the alternative parameters, and hence the non-centrality parameter as quadratic in the alternative parameters.

The solid line is almost exactly horizontal at the target power. The discrepancy from horizontal arises from the error in approximating (4.27) by (4.30). For small sample sizes (that is, group sizes of less than 25, or group sizes of less than 75), (4.28) is not sufficiently accurate; for larger sample sizes it should be accurate enough.

All curves in Figure 4.3 use the approximate critical value in (4.18).

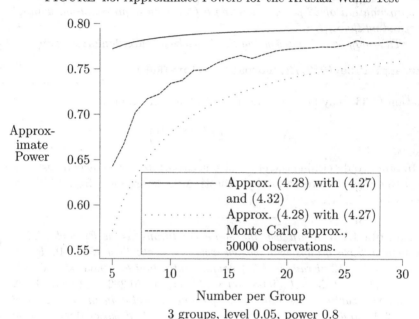

FIGURE 4.3: Approximate Powers for the Kruskal-Wallis Test

3 groups, level 0.05, power 0.8

4.8 Efficiency Calculations

Power calculations for one-dimensional alternative hypotheses made use of (2.22), applying a Gaussian approximation with exact values for means under the null and alternatives, and approximating the variance under the alternative by the variance under the null. Efficiency calculations of §2.4 and §3.8 approximated means at the alternative linearly using the derivative of the mean function at the null. Tests for ordered alternatives from this chapter will be addressed similarly.

4.8.1 Ordered Alternatives

Consider first the one-sided Jonckheere-Terpstra test of level α. Let $T_J = J/N^2$. In this case, the subscript J represents a label, and not an index. Denote the critical value by t_J°, satisfying $P_{\theta^\circ}[T_J \geq t_J^\circ] = 1 - \alpha$.

As in (4.23), reduce the alternative hypothesis to a single dimension by letting the alternative parameter vector be a fixed vector times a multiplier Δ. The power function $\varpi_{J,n}(\Delta) = P_{\theta^A}[T_J \geq t_J^\circ]$ satisfies (2.15), (2.19), and (2.21), and hence the efficiency tools for one-dimensional hypotheses developed in §2.4.2 may be used. Expressing $\mu_J(\Delta)$ as a Taylor series with constant and linear terms,

$$\mu_J(\Delta) \approx \sum_{i=1}^{K-1} \sum_{j=i+1}^{K} \lambda_i \lambda_j (\kappa^\circ + \kappa'[\theta_j^A - \theta_i^A])$$

for $\lambda_i = M_i/N$, where again κ° is the common value of κ_{jk} under the null hypothesis, and κ' is the derivative of the probability in (3.25), as a function of the location shift between two groups, evaluated at the null hypothesis, and calculated for various examples in §3.8.2. Hence

$$\mu'_J(0) = \sum_{i=1}^{K-1} \sum_{j=i+1}^{K} \lambda_i \lambda_j \kappa'[\theta_j^\dagger - \theta_i^\dagger].$$

Recall that κ_{ij} depended on two group indices i and j only because the locations were potentially shifted relative to one another; the value κ° and its derivative κ' are evaluated at the null hypothesis of equality of distributions, and hence do not depend on the indices. Furthermore, from (4.21), $\text{Var}[T_J] \approx \frac{1}{36}\left[1 - \sum_{k=1}^{K} \lambda_k^3\right]/N$. Consider the simple case in which $\lambda_k = 1/K$ for all k, and in which $\theta_j^\dagger - \theta_i^\dagger = (j - i)$. Then $\mu'_J(0) = \kappa'(K^2 - 1)/6K$, $\text{Var}[T_J] \approx \frac{1}{36}\left[1 - 1/K^2\right]/N$, and

$$e_J = \kappa' \frac{(K^2-1)/6K}{\sqrt{\frac{1}{36}[1 - 1/K^2]}} = \frac{\kappa'(K^2-1)}{\sqrt{K^2-1}} = \kappa'\sqrt{K^2-1}.$$

The efficacy of the Gaussian-theory contrast, from (4.6), is $\sqrt{K^2-1}/(2\sqrt{3}\varepsilon)$. Hence the asymptotic relative efficiency of the Jonckheere-Terpstra test to the contrast of means is

$$(\kappa')^2(12\varepsilon^2). \tag{4.35}$$

This is the same as the asymptotic relative efficiency in the two-sample case in Table 3.6.

4.8.2 Unordered Alternatives

While techniques exist to consider transform non-central chi-square statistics to an approximate Gaussian distribution (Sankaran, 1963), the most direct approach to comparing efficiency of various tests based on approximately χ^2 statistics is to compare approximate sample sizes for a fixed level and power, as in (4.33). Prentice (1979) does this using ratios of the non-centrality parameter. For test j, let N_j, be the sample size needed to provide power $1 - \beta$ for the test of level j against alternative θ^A.

Calculations (4.27), and (4.30) through (4.34) were motivated specifically for the Kruskal-Wallis test, using Mann-Whitney sums (4.26), but hold more broadly for any group summary statistics replacing (4.26), so long as $E_0[T_k] \approx M_k \sum_{l=1, l \neq k}^{K} M_l \kappa^\circ$ for some differentiable κ°, and as long as (4.25), with the new definition of T_k, is approximately χ^2_{K-1}. In particular, a rescaled version of the F-test statistic (4.3) $(K-1)W_A = \sum_{k=1}^{K} M_k(\bar{X}_{k.} - \bar{X}_{..})^2/\tilde{\sigma}^2$ is such a test, with ψ the variance of the X_{kj}, $T_k = \bar{X}_{k.} - \bar{X}_{..}$, $\kappa^\circ = 0$, and $\kappa' = 1$, approximated as the χ^2 statistic by treating the variance as known. Suppose that the value of ζ^2 remains unchanged for two such tests, and suppose these tests have values of κ' at the null hypothesis distinguished by indices (for concreteness, label these as κ'_H and $\kappa'_A = 1$ for the correctly-scaled Kruskal-Wallis and F tests respectively), and similarly having the norming factors ψ distinguished by indices (again, $\psi_H = 1/\sqrt{12}$ and $\psi_A = \varepsilon^2$). Equate the alternatives in (4.34), to obtain

$$\psi_H \sqrt{\xi} / \left(\kappa'_H \zeta N_H^{1/2} \right) \approx \psi_A \sqrt{\xi} / \left(\kappa'_A \zeta N_A^{1/2} \right).$$

The ratio of sample sizes needed for approximately the same power for the same alternative from the F and Kruskal-Wallis tests is approximately

$$N_A/N_H \approx (\psi_A^2/\psi_H^2) \left((\kappa'_H)^2/(\kappa'_A)^2 \right) = 12\varepsilon^2 (\kappa'_H)^2,$$

which is the same as in the unordered case (4.35) and in the two-sample case.

4.9 Exercises

1. The data set

    http://ftp.uni-bayreuth.de/math/statlib/datasets/federalistpapers.txt

    gives data from an analysis of a series of documents. The first column gives document number, the second gives the name of a text file, the third gives a group to which the text is assigned, the fourth represents a measure of the use of first person in the text, the fifth presents a measure of inner thinking, the sixth presents a measure of positivity, and the seventh presents a measure of negativity. There are other columns that you can ignore. (The version at Statlib,

Exercises 93

above, has odd line breaks. A reformatted version can be found at stat.rutgers.edu/home/kolassa/Data/federalistpapers.txt).

a. Test the null hypothesis that negativity is equally distributed across the groups using a Kruskal-Wallis test.

b. Test at $\alpha = .05$ the pairwise comparisons for negativity between groups using the Bonferroni adjustment, and repeat for Tukey's HSD.

2. The data set

   http://ftp.uni-bayreuth.de/math/statlib/datasets/Plasma_Retinol

   gives the results of a study on concentrations of certain nutrients among potential cancer patients. This set gives data relating various quantities, including smoking status (1 never, 2 former, 3 current) in column 3 and beta plasma in column 13. Perform a nonparametric test to investigate an ordered effect of smoking status on beta plasma.

3. Consider using a Kruskal-Wallis test of level 0.05, testing for equality of distribution for four groups of size 15 each. Consider the alternative that two of these groups have observations centered at zero, and the other two have observations centered at 1, with all observations from a Cauchy distribution.

   a. Approximate the power for this test to successfully detect the above group differences.

   b. Check the quality of this approximation via simulation.

   c. Approximate the size in each group if a new study were planned, with four equally-sized groups, and with the groups centered as above, with a target power of 90%.

4. The data at

   http://lib.stat.cmu.edu/datasets/CPS_85_Wages

   reflects wages from 1985. The first 42 lines of this file contain a description of the data set, and an explanation of variables; delete these lines first, or skip them when you read the file. All fields are numeric. The tenth field is sector, and the sixth field is hourly wage; you can skip everything else. Test for a difference in wage between various sector groups, using a Kruskal–Wallis test.

5

Group Differences with Blocking

This chapter concerns testing and estimation of differences in location among multiple groups, as raised in the previous chapter, in the presence of blocking. Techniques useful when data are known to be approximately Gaussian are first reviewed, for comparison purposes. The simplest example of blocking is that of paired observations; paired observations are considered after Gaussian techniques. More general multi-group inference questions are addressed later.

5.1 Gaussian Theory Approaches

Techniques appropriate when observations are well-approximated by a Gaussian distribution are well-established. These are first reviewed to give a point of departure for nonparametric techniques, which are the focus of this volume.

5.1.1 Paired Comparisons

Suppose pairs of values (X_{i1}, X_{i2}) are observed on n subjects, indexed by i, and suppose further that (X_{i1}, X_{i2}) is independent of (X_{j1}, X_{j2}) for $i, j \in \{1, \ldots, n\}$, $i \neq j$, and that all of the vectors (X_{i1}, X_{i2}) have the same distribution. Under the assumption that the observations have a Gaussian distribution, one often calculates differences $Z_i = X_{i1} - X_{i2}$, and applies either the testing procedure of §2.1.2, or the associated confidence interval procedure. Such a test is called a paired t-test.

5.1.2 Multiple Group Comparisons

Suppose one observes independent random variable X_{kli}, with $k \in \{1, \cdots, K\}$, $l \in \{1, \cdots, L\}$, and $i \in \{1, \cdots, M_{kl}\}$. Here index k represents treatment, and index l represents block. A test of equality of treatment means is desired, without an assumption about the effect of block. When these variables are Gaussian, and have equal variance, one can construct a statistic analogous to the F test (4.3), in which the numerator is a quadratic form in differences between the means specific to group, $\bar{X}_{k..} = \sum_{l=1}^{L} \sum_{i=1}^{M_{kl}} X_{kli} / \sum_{l=1}^{L} n_{kl}$, and the overall mean. Formulas are available for the statistic in closed form only when the replicate numbers satisfy

$$M_{kl} = M_{k.}M_{.l}/M_{..} \tag{5.1}$$

for $M_{k.} = \sum_{l=1}^{L} M_{kl}$, $M_{.l} = \sum_{k=1}^{K} M_{kl}$, and $M_{..} = \sum_{k=1}^{K} \sum_{l=1}^{L} M_{kl}$.

5.2 Nonparametric Paired Comparisons

Suppose pairs of values (X_{i1}, X_{i2}) are observed on n subjects, indexed by i, and suppose further that (X_{i1}, X_{i2}) is independent of (X_{j1}, X_{j2}) for $i, j \in \{1, \ldots, n\}$, $i \neq j$, and that all of the vectors (X_{i1}, X_{i2}) have the same distribution. Consider the null hypothesis that the marginal distribution of $\{X_i\}$ is the same as that of $\{Y_i\}$, versus the alternative hypothesis that the distributions are different. This null hypothesis often, but not always, implies that the differences $Z_i = X_{i1} - X_{i2}$ have zero location. One might test these hypotheses by calculating the difference Z_i between values, and apply the one-sample technique of Chapter 2, the sign test; this reduces the problem to one already solved.

After applying the differencing operation, one might expect the differences to be more symmetric than either of the two original variables, and one might exploit this symmetry. To derive a test under the assumption of symmetry, Let R_j be the rank of $|Z_j|$ among all absolute values. Let

$$S_j = \begin{cases} 1 & \text{if } i\, j \text{ is positive} \\ 0 & \text{if } i\, j \text{ is negative} \end{cases}.$$

Define the Wilcoxon signed-rank test in terms of the statistic

$$T_{SR} = \sum R_j S_j. \tag{5.2}$$

Under the null hypothesis that the distribution of the X_i is the same as the distribution of the Y_i, and again, assuming symmetry of the differences, then $(, S_j,)$ and $(, R_j,)$ are independent random vectors, because S_j and $|X_j|$ are pairwise independent under H_0.

Components of the random vector $(R_1, \ldots, R_n)$ are dependent, and hence calculation of the variance from T_{SR} via (5.2) requires calculation of the sum of identically distributed but not independent random variables. An alternative formulation, as the sum of independent but not identically distributed random variables, will prove more tractable. Let

$$V_j = \begin{cases} 1 & \text{if the item whose absolute value is ranked } j \text{ is positive} \\ 0 & \text{if the item whose absolute value is ranked } j \text{ is negative} \end{cases}.$$

Nonparametric Paired Comparisons

Hence $T_{SR} = \sum jV_j$, and the null expectation and variance are

$$\text{E}_0[T_{SR}] = \sum j\text{E}_0[V_j] = n(n+1)/4 \qquad (5.3)$$

and

$$\text{Var}_0[T_{SR}] = \sum j^2 \text{Var}_0[V_j] = \sum j^2/4 = n(2n+1)(n+1)/24. \qquad (5.4)$$

One can also calculate exact probabilities for T_{SR} recursively, as one could for the two-sample statistic. There are 2^n ways to assign signs to ranks $1, \ldots, n$. Let $f(t, n)$ be the number of such assignments yielding $T_{SR} = t$ with n observations. Again, as in §3.4.1, summing the counts for shorter random vectors with alternative final values,

$$f(t,n) = \begin{cases} 0 & \text{for } t < 0 \text{ or } t > n(n+1)/2 \\ 1 & \text{if } n = 1 \text{ and } t \in \{0, 1\} \\ f(t, n-1) + f(t-n, n-1) & \text{otherwise} \end{cases}.$$

This provides a recursion that can be used to calculate exact p-values.

Example 5.2.1 *Consider data calculated on the size of brains of twins (Tramo et al., 1998). This data set from*

http://lib.stat.cmu.edu/datasets/IQ_Brain_Size

contains data on 10 sets of twins. Each child is represented by a separate line in the data file, for a total of 20 lines. We investigate whether brain volume (in field 9) is influenced by birth order (in field 4). Brain volumes for the first and second child, and their difference, are given in Table 5.1. The rank sum statistic is $3+4+10+8+7 = 32$, the null expected rank sum is $10 \times 11/4 = 27.5$, the null variance is $10 \times 21 \times 11/24 = 96.25$, and so the two-sided p-value is $2 \times \Phi(-(32-.5-27.5)/\sqrt{96.25}) = 2 \times \Phi(-0.408) = .683$. There is no evidence that twin brain volume differs by birth order. This calculation might also have been done in R using

```
twinbrain<-as.data.frame(scan("IQ_Brain_Size",
    what=list(CCMIDSA=0,FIQ=0,HC=0,ORDER=0,PAIR=0,SEX=0,
    TOTSA=0, TOTVOL=0,WEIGHT=0),skip=27,nmax=20))
fir<-twinbrain[twinbrain$ORDER==1,]
fir$v1<-fir$TOTVOL
sec<-twinbrain[twinbrain$ORDER==2,]
sec$v2<-sec$TOTVOL
brainpairs<-merge(fir,sec,by="PAIR")[,c("v1","v2")]
brainpairs$diff<-brainpairs$v2-brainpairs$v1
wilcox.test(brainpairs$diff)
```

giving an exact p-value of 0.695. Compare these to the results of the sign test and t-test:

TABLE 5.1: Twin brain volume

Pair	1	2	3	4	5	5	7	8	9	10
First	1005	1035	1281	1051	1034	1079	1104	1439	1029	1160
Second	963	1027	1272	1079	1070	1173	1067	1347	1100	1204
Diff.	-42	-8	-9	28	36	94	-37	-92	71	44
Rank	6	1	2	3	4	10	5	9	8	7

```
library(BSDA)#Need for sign test.
SIGN.test(brainpairs$diff)
t.test(brainpairs$diff)
```

giving p-values of 1 and 0.647 respectively.

As with the extension from two-sample testing to multi-sample testing referred to in §3.4.2, one can extend the other rank-based modifications of §3.4.2 and §3.6 to the blocking context as well. Ties may be handled by using mean ranks, and testing against a specific distribution for the alternative may be tuned using the appropriate scores. The more general score statistic for paired data is $T_{GP} = \sum_j a_j S_j$; in this case, $\mathrm{E}[T_{GP}] = \sum_j a_j/2$, and $\mathrm{Var}[T_{GP}] = \sum_j a_j^2/4$.

Example 5.2.2 *Perform the asymptotic score tests on the brain volume differences of Example 5.2.1.*

```
library(MultNonParam)
cat("Asymptotic test using normal scores\n")
brainpairs$normalscores<-qqnorm(seq(length(
   brainpairs$diff)),plot.it=F)$x
symscorestat(brainpairs$diff,brainpairs$normalscores)
cat("Asymptotic test using savage scores\n")
brainpairs$savagescores<-cumsum(
   1/rev(seq(length(brainpairs$diff))))
symscorestat(brainpairs$diff,brainpairs$savagescores)
```

giving p-values of 0.863 and 0.730 respectively.

Permutation testing can also be done, using raw data values as scores.

This procedure uses the logic that X_j and Y_j having the same marginal distribution implies $Y_j - X_j$ has a symmetric distribution. Joint distributions exist for which this is not true, but these examples tend to be contrived.

5.2.1 Estimating the Population Median Difference

Estimation of the median of the observations, which is their point of symmetry, based on inversion of the Wilcoxon signed-rank statistic mirrors that based

Nonparametric Paired Comparisons

on inversion of the Mann-Whitney-Wilcoxon statistic of §3.10. Let $T_{SR}(\theta)$ be the Wilcoxon signed-rank statistic calculated from the data set $Z_j(\theta) = Z_j - \theta = Y_j - X_j - \theta$, after ranking the $Z_j(\theta)$ by their absolute values, and summing the ranks with positive values of $Z_j(\theta)$. Estimate θ as the quantity $\hat{\theta}$ equating T_{SR} to the median of its sampling distribution; that is, $\hat{\theta}$ satisfies $T_{SR}(\hat{\theta}) = n(n+1)/4$.

After sorting the values Z_j, denote value at position j in the ordered list by the order statistic $Z_{(j)}$. One can express $\hat{\theta}$ in terms of $Z_{(j)}$, by considering the behavior of $T_{SR}(\theta)$ as θ varies. For $\theta > Z_{(n)}$, $T_{SR}(\theta) = 0$. When θ decreases past $Z_{(n)}$, the $Z_i - \theta$ with the lowest absolute value switches from negative to positive, and $T_{SR}(\theta)$ moves up to 1. When θ decreases past $(Z_{(n)} + Z_{(n-1)})/2$, then $Z_{(n)} - \theta$ goes from having the smallest absolute value to having the second lowest absolute value, and T_{SR} goes to 2. The next jump in T_{SR} occurs if one more observation becomes positive (at $Z_{(n-1)}$) or if the absolute value of the lowest shifted observation passes the absolute value of the next observation (at $Z_{(n)} + Z_{(n-2)}/2$).

Generally, the jumps happen at averages of two observations, including averages of an observation with itself. These averages are called Walsh averages. These play the same role as differences in the two-sample case. First, note that T_{SR} is the number of positive Walsh averages. This can be seen by letting $Z_{[i]}$ be observation ordered by absolute value, and letting $W_{ij} = (Z_{[i]} + Z_{[j]})/2$ for $i \leq j$. Suppose that $Z_{[j]} > 0$. Then $Z_{[i]} + Z_{[j]} > 0$ for $i < j$, and all $W_{ij} > 0$ for $i < j$, and $W_{jj} > 0$. On the other hand, if $Z_{[j]} < 0$, then $Z_{[i]} + Z_{[j]} < 0$ for $i < j$, and then all $W_{ij} < 0$ for $i < j$, and $W_{jj} < 0$.

So $\hat{\theta}$ has half of the Walsh averages below it, and half above, and $\hat{\theta}$ is the median of Walsh averages. This estimator is given by Hodges and Lehmann (1963), in the same paper giving the analogous estimator for the one-sample symmetric context of §3.10, and is called the Hodges-Lehmann estimator.

Example 5.2.3 *Walsh averages may be extracted in R using*

```
aves<-outer(brainpairs$diff,brainpairs$diff,"+")/2
sort(aves[upper.tri(aves,diag=TRUE)])
```

to obtain the $10 \times (10+1)/2 = 55$ *pairwise averages*

```
-92.0 -67.0 -64.5 -50.5 -50.0 -42.0 -39.5 -37.0 -32.0 -28.0
-25.5 -25.0 -24.0 -23.0 -22.5 -10.5  -9.0  -8.5  -8.0  -7.0
 -4.5  -3.0  -0.5   1.0   1.0   3.5   9.5  10.0  13.5  14.0
 14.5  17.0  17.5  18.0  26.0  28.0  28.5  31.0  31.5  32.0
 36.0  36.0  40.0  42.5  43.0  44.0  49.5  53.5  57.5  61.0
 65.0  69.0  71.0  82.5  94.0
```

Their median is observation 28, which is 10.0. Estimate the median difference as 10.0.

5.2.2 Confidence Intervals

Confidence intervals may also be constructed using the device of (1.16), similarly as with the one-sample location interval construction of §2.3.3 and the two-sample location shift interval construction of §3.10. Let $\tilde{W}_j$ be the ordered Walsh averages. Find the largest t_l such that $P[T_{SR} < t_l] < \alpha/2$; then t_l is the $\alpha/2$ quantile of the distribution of the Signed Rank statistic. Using (5.3) and (5.4), one might approximate the critical value using a Gaussian approximation $t_l \approx n(n+1)/2 - z_{\alpha/2}\sqrt{n(2n+1)(n+1)/24}$; note that this approximation uses the distribution of the signed-rank statistic, which does not depend on the distribution of the underlying data as long as symmetry holds. Recall that z_β is the $1-\beta$ quantile of the standard Gaussian distribution, for any $\beta \in (0,1)$; in particular, $z_{\alpha/2}$ is positive for any $\alpha \in (0,1)$, and $z_{0.05/2} = 1.96$. By symmetry, $P[T_{SR} \geq t_u] \leq \alpha/2$ for $t_u = n(n-1)/2 - t_l + 1$.

As noted above, $T_{SR}(\theta)$ jumps by one each time θ passes a Walsh average. Hence the confidence interval is

$$(\tilde{W}_{n(n-1)/2+1-t_l}, \tilde{W}_{n(n-1)/2+1-t_u}). \tag{5.5}$$

By symmetry, this interval is $(\tilde{W}_{t_l}, \tilde{W}_{t_u})$. See Figure 5.1.

FIGURE 5.1: Construction of Median Estimator

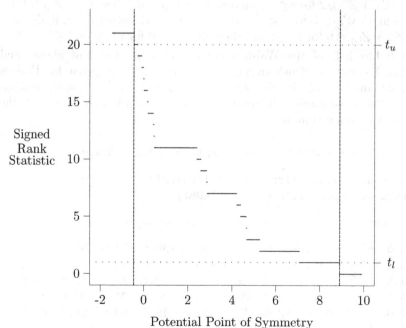

Hypothetical Data Set with 6 Observations

Nonparametric Paired Comparisons 101

> **Example 5.2.4** *Refer again to the brain volume data of Example 5.2.1. Find a 95% confidence interval for the difference in brain volume. The 0.025 quantile of the Wilcoxon signed-rank statistic with 10 observations is $t° = 9$; this can be calculated from R using*
>
> ```
> qsignrank(0.025, 10)
> ```
>
> *and confidence interval endpoints are observations 9 and 55+1-9=47. As tabulated above, Walsh average 9 and 47 are -32.0 and 49.5 respectively. Hence the confidence interval is (-32.0, 49.5). This might have been calculated directly in R using*
>
> ```
> wilcox.test(brainpairs$diff,conf.int=TRUE)
> ```

Similar techniques to those of this section were used in §2.3.3 to give confidence intervals for the median, and in §3.10 to give confidence intervals for median differences. In each case, a statistic dependent on the unknown parameter was constructed, and estimates and confidence intervals were constructed by finding values of the parameter equating the statistic to appropriate quantiles. However, the treatment of this section and that of §2.3.3 differ, in that the resulting statistic in this section is non-increasing in θ in this section, while in §2.3.3 it was non-decreasing. The increasing parameterization of §2.3.3 was necessary to allow the same construction to be used in §2.3.4, when quantiles other than the median were estimated; these quantiles estimates are an increasing function of each observation separately. A parallel construction might have been used in the current section, by inverting the statistic formed by summing ranks of negative observations; as this definition runs contrary to the common definition of the Wilcoxon Signed Rank statistic, it was avoided.

5.2.3 Signed-Rank Statistic Alternative Distribution

Consider alternative hypotheses H_A, specifying that the median of the distribution of differences is θ. Hence

$$\begin{aligned} E_\theta[T] &= \sum_{i=1}^{n}\sum_{j=1}^{i} P_\theta\left[(Z_i + Z_j - 2\theta) \leq 0\right] \\ &= n(n-1)P[Z_1 + Z_2 \leq 2\theta]/2 + nP[Z_1 \leq \theta]. \end{aligned}$$

To apply asymptotic relative efficiency calculations, scale the test statistic to have an expectation that varies with the parameter value specifying the null hypothesis, and approximately independent of sample size, as in (2.15), by switching to $S = 2T_{SR}/(n(n-1))$. In this case, $\mu(\theta) \approx P_0[Z_1 + Z_2 \geq 2\theta]$, and the statistic variance times the sample size is

$$\sigma^2(0) = 4 \times n \times n(n+1)(2n+1)/(n^2(n+1)^2 24) \approx 1/6,$$

to match the notation of (2.15) and (2.19). Note that $\mu'(0)$ is now twice the value from the Mann-Whitney-Wilcoxon statistic, for distributions symmetric about 0. This will be used for asymptotic relative efficiency calculations in the exercises.

5.3 Two-Way Non-Parametric Analysis of Variance

Consider the same data description as in §5.1.2: one observes random variable X_{kli}, with $k \in \{1, \cdots, K\}$, $l \in \{1, \cdots, L\}$, and $i \in \{1, \cdots, M_{kl}\}$. Suppose the distribution of X_{kli} are independent, with a distribution that is allowed to depend on k and l; that is, $X_{kli} \sim F_{kl}$. We wish to test the null hypothesis that treatment has no effect, while allowing different blocks to behave differently, versus the alternative that the distribution depends on treatment as well. That is, the null hypothesis is that F_{kl} does not depend on k, although it may depend on l, and the alternative hypothesis is that for some l, some j and k, and some x, $F_{kl}(x) \neq F_{jl}(x)$. However, the test to be constructed will not have reasonable power unless this difference is more systematic; that is, consider alternatives for which for each pair of treatments j and k, either $F_{jl}(x) \leq F_{jl}(x)$ for all x and all l, or $F_{jl}(x) \geq F_{jl}(x)$ for all x and all l. The direction of the effect must be constant across blocks. Heuristically, treat k as indexing treatment and l as indexing a blocking factor.

In order to build a statistic respecting treatment order across blocks, rank the observations within blocks; that is, let R_{kli} represent the rank of X_{kli} within $X_{1l1}, \ldots, X_{KlM_{Kl}}$, and sum them separately by block and treatment.

5.3.1 Distribution of Rank Sums

In order to use these ranks to detect treatment differences that are in a consistent direction across blocks, one should consolidate rank sums over blocks before comparing across treatments. Let $R_{k..} = \sum_{l=1}^{L} \sum_{i=1}^{M_{kl}} R_{kli}$ and $\bar{R}_{k..} = R_{k..} / \sum_{l=1}^{L} M_{kl}$. Consolidation via averaging, rather than via summing, implies a choice in weighting the contributions of the various blocks.

Moments of these rank sums may be calculated as an extension of (3.22). Under the null hypothesis of no treatment effect, the expectation of one rank is

$$\mathrm{E}\left[R_{kli}\right] = \left(\sum_{j=1}^{K} M_{jl} + 1\right)/2,$$

and hence

$$\mathrm{E}\left[R_{k..}\right] = \sum_{l=1}^{L} \left[M_{kl}\left(\sum_{j=1}^{K} M_{jl} + 1\right)/2\right]. \quad (5.6)$$

Variances of rank sums depend on the covariance structure of the ranks. Ranks that make up each sum within a block are independent, but sums of ranks across blocks are dependent. Within a treatment-block cell, $\text{Var}[R_{kl.}]$ is the same as for the Mann-Whitney-Wilcoxon statistic:

$$\text{Var}[R_{kl.}] = M_{kl} \left(\sum_{j \neq k} M_{jl} \right) \left(\sum_{j=1}^{K} M_{jl} + 1 \right) / 12.$$

By independence across blocks,

$$\text{Var}[R_{k..}] = \sum_{l=1}^{L} \left\{ M_{kl} \left(\sum_{j \neq k} M_{jl} \right) \left(\sum_{j=1}^{K} M_{jl} + 1 \right) / 12 \right\}. \quad (5.7)$$

Covariances may be calculated by comparing the variance of the sum to the sum of the variances, as in (4.12), to obtain

$$\text{Cov}[R_{kl.}, R_{ml.}] = -[M_{kl} M_{ml}] \left(\sum_{j=1}^{K} M_{jl} + 1 \right) / 12.$$

Since blocks are ranked independently, variances and covariances for rank sums add across blocks:

$$\text{Cov}[R_{k..}, R_{m..}] = - \sum_{l=1}^{L} [M_{kl} M_{ml}] \left(\sum_{j=1}^{K} M_{jl} + 1 \right) / 12. \quad (5.8)$$

5.4 A Generalization of the Test of Friedman

From these rank sums one might calculate separate Kruskal-Wallis statistics for each block, and, since statistics from separate blocks are independent, one might add these Kruskal-Wallis statistics to get $\chi^2_{L(K-1)}$ degree of freedom statistic under the null hypothesis. Such a statistic, however, is equally sensitive to deviations from the null hypothesis in opposite directions in different blocks, and so has low power against all alternatives. Interesting alternative hypotheses involve treatment distributions in different blocks ordered in the same way.

As with the Kruskal-Wallis test, a statistic is constructed by setting $\mathbf{Y} = (R_{1..} - E_0[R_{1..}], \ldots, R_{K-1..} - E_0[R_{K-1..}])^\top$, using (5.6), calculating $\text{Var}_0[\mathbf{Y}]$ from (5.7) and (5.8), and using

$$W_F = \mathbf{Y}^\top \text{Var}_0[\mathbf{Y}]^{-1} \mathbf{Y} \quad (5.9)$$

as the test statistic.

5.4.1 The Balanced Case

The inverse $\text{Var}_0 [\boldsymbol{Y}]^{-1}$ is tractable in closed form under (5.1), including the special case with balanced replicates.

Friedman (1937) addresses the case in which $M_{kl} = 1 \forall k, l$. The generalization to the balanced case with replicates is trivial, and this section assumes $M_{kl} = M$ for all k, l. Here $\text{E}\left[\bar{R}_{k..}\right] = (KM+1)/2$, and

$$\text{Var}\left[\bar{R}_{k..}\right] = (K-1)(KM+1)/[12L].$$

Covariances between sets of rank means are

$$\text{Cov}\left[\bar{R}_{k..}, \bar{R}_{m..}\right] = -(KM+1)/[12L]$$

for $k \neq m$. In this balanced case, the test statistic (5.9) can be shown to equal the sum of squares of ranks away from their average value per treatment group:

$$W_F = 12L \sum_{k=1}^{K} [\bar{R}_{k..} - (MK+1)/2]^2 / [K(KM+1)].$$

This demonstration is as for the Kruskal-Wallis test in §4.2.2. The limiting distribution of W_F is also demonstrated by observing the parallelism between this covariance structure and the covariance structure of rank sums in §4.2.2, since the covariance structure of $\boldsymbol{Y} = (\bar{R}_{1..}, \ldots, \bar{R}_{K-1..})/\sqrt{K}$ has the same form as (4.12), for some vector $\boldsymbol{\nu}$ and constant ω. A relationship like that of §4.2.2 can then be used to demonstrate that W_F is approximately χ^2_{K-1}. Again, some condition on the sample sizes being large enough to ensure that the rank averages are approximately multivariate Gaussian is required; for example, in the case with the same number of replicates per treatment-block combination, having the number of blocks going to infinity suffices, and in the balanced case with a fixed number of blocks, having the number of observations in each block-treatment combination go to infinity suffices. This requirement is in addition to (5.1) needed to find a generalized inverse for the variance matrix in closed form, and hence to exhibit the test statistic as a sum of squares.

Example 5.4.1 *Friedman (1937) provides data on variability of expenses in various categories for families of different income levels. This data is provided in Table 5.2, and at*

http://stat.rutgers.edu/home/kolassa/Data/friedman.dat .

This may be read into R using

```
expensesd<-as.data.frame(scan("friedman.dat",
    what=list(cat="",g1=0,g2=0,g3=0,g4=0,g5=0,g6=0,g7=0)))
```

A Generalization of the Test of Friedman

Here $L = 14$, $M = 1$, and $K = 7$. Ranking within groups (for example, in R, using

```
expenserank<-t(apply(as.matrix(expensesd[,-1]),1,rank))
rownames(expenserank)<-expensesd[[1]]
```

gives the table of ranks

	g1	g2	g3	g4	g5	g6	g7
Housing	5	1	3	2	4	6	7
Operations	1	3	4	6	2	5	7
Food	1	2	7	3	5	4	6
Clothing	1	3	2	4	5	6	7
Furnishings	2	1	6	3	7	5	4
Transportation	1	2	3	6	5	4	7
Recreation	1	2	3	4	7	5	6
Personal	1	2	3	6	4	7	5
Medical	1	2	4	5	7	3	6
Education	1	2	4	5	3	6	7
Community	1	5	2	3	7	6	4
Vocation	1	5	2	4	3	6	7
Gifts	1	2	3	4	5	6	7
Other	5	4	7	2	6	1	3

Rank sums are $23, 36, 53, 57, 70, 70, 83$, *and group means (for example, via*

```
apply(expenserank,2,mean)
```

gives $1.643, 2.571, 3.786, 4.071, 5.000, 5.000, 5.929$. *Subtracting the overall mean,* 4, *squaring, and adding, gives* 13.367. *Multiplying by* $12L/(K(K+1)) = 12 \times 14/(7 \times 8) = 3$ *gives the statistic value* $W_F = 40.10$. *Comparing to the* χ_6^2 *distribution gives a p-value of* 4.35×10^{-7}. *This might also have been done entirely in R using*

```
friedman.test(as.matrix(expensesd[,-1]))
```

The development of Friedman's test in the two-way layout remarkably preceded analogous testing in one-way layouts. When $K = 2$ and M_{kl} are all 1, Friedman's test is equivalent to the sign test applied to differences within a block.

5.4.2 The Unbalanced Case

This analysis fails in the unbalanced case, because no expression for the variance as simple as (4.12) holds, and so the variance matrix cannot be inverted in closed form. Benard and Elteren (1953) present a numerical method that uses the rule of Cramer to produce the matrix product involved in T. Pren-

TABLE 5.2: Expense variability by income group, from Friedman (1937)

Category	Gr. 1	Gr. 2	Gr. 3	Gr. 4	Gr. 5	Gr. 6	Gr. 7
Housing	103.30	68.42	89.53	77.94	100.00	108.20	184.90
Operations	42.19	44.31	60.91	73.90	43.87	61.74	102.30
Food	71.27	81.88	100.71	86.52	100.30	90.75	100.60
Clothing	37.59	60.05	56.97	60.79	71.82	83.04	117.10
Furnishings	58.31	52.73	96.04	60.42	104.33	89.78	85.77
Transport	46.27	82.18	129.80	181.00	172.33	164.80	246.80
Recreation	19.00	23.07	38.70	45.81	59.03	50.69	55.18
Personal	8.31	8.43	9.16	14.28	10.63	15.84	12.50
Medical	20.15	33.48	60.08	69.35	114.34	45.28	101.60
Education	3.16	4.12	12.73	18.95	8.89	41.52	66.33
Community	4.12	18.87	8.54	12.92	25.30	19.85	16.76
Vocation	7.68	11.18	10.44	10.95	10.54	13.96	14.39
Gifts	5.29	10.91	11.22	25.26	42.25	48.80	69.38
Other	6.00	5.57	22.23	2.45	6.24	1.00	4.00

tice (1979) performs these calculations in somewhat more generality, and the general unbalanced two-way test is generally called the Prentice test. Skillings and Mack (1981) address this question using explicit numerical matrix inversion.

Example 5.4.2 *Consider data on weight gain among chickens, given by Cox and Snell (1981, Example K), and at*

http://stat.rutgers.edu/home/kolassa/Data/chicken.dat .

Weight gain after 16 weeks, protein level, protein source, and fish soluble level. The dependence of weight gain on protein source might be graphically depicted using box plots (Figure 5.2):

```
temp1<-temp<-as.data.frame(scan("chicken.dat",what=
   list(source="", lev=0,fish=0,weight=0,w2=0)))
temp1$weight<-temp1$w2;temp$h<-0;temp1$h<-1
temp$w2<-NULL;temp1$w2<-NULL
chicken<-rbind(temp,temp1)
attach(chicken)
boxplot(split(weight,source),horizontal=TRUE,ylab="g.",
   main="Weight gain for Chickens", xlab="Protein Source")
detach(chicken)
```

Hence blocking on source is justified. Test for dependence of weight gain on protein level, blocking on source. Perform these calculations in R using

```
library(muStat)#Need for Prentice test.
```

```
attach(chicken)
prentice.test(weight,source,blocks=lev)
detach(chicken)
```

to obtain the p-value 0.1336.

FIGURE 5.2: Weight Gain for Chickens

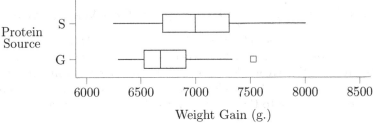

5.5 Multiple Comparisons and Scoring

One might adjust for multiple comparisons in the presence of blocking in the same way as was done without blocking, as discussed in §4.5. That is, one might use the variances and covariances to show that under the null hypothesis, $(\bar{R}_{i.} - \bar{R}_{j.})/\sqrt{K(LK+1)/6}$ is approximately Gaussian, and employ the method of Bonferroni, Fisher's LSD, or Tukey's HSD.

One might also use scoring methods, including normal and Savage scores, as before. Recall also that treating ranks as general scores, with tied observations given average ranks, makes tie handling automatic.

One might also use scores that are data values. In this case, the test statistic is similar to that of the Gaussian-theory F test, but the reference distribution is generated from all possible permutations of the data within blocks and among treatments.

Example 5.5.1 *Refer again to Example 5.4.2. Test dependence of weight gain on protein source, blocking on level. There are eight observations at each protein level. Cycle through all $\binom{8}{4}^3 = 34300$ ways to associate source labels within block, and compute the between sum of squares each time. The p-value is the number of rearrangements having this statistic meet or exceed the observed.*

```
library(MultNonParam)
#aov.P requires data sorted by block.  Put block ends as the
#third argument.
```

```
chicken<-chicken[order(chicken$lev),]
aov.P(chicken$weight,as.numeric(as.factor(chicken$source)),
  c(8,16,24))
```
The p-value is 0.182.

5.6 Tests for a Putative Ordering in Two-Way Layouts

Recall Friedman's statistic of (5.9), $W_F \propto \sum_{k=1}^{K}[\bar{R}_{k..} - 1/2(MK+1)]^2$, where $\bar{R}_{k..}$ are mean rankings for treatment k when observations are ranked within blocks. The statistic W_F treats deviation of any treatment in any direction similarly. If one wants to look specifically for alternatives suspected *a priori* of being ordered according to the index k, in the balanced case use instead

$$T_L = \sum_{k=1}^{K} k R_{k..}. \tag{5.10}$$

This test was proposed by Page (1963) in the balanced case with one replicate per block-treatment pair, and is called Page's test.

For more general replication patterns, the null expectation and variance for the statistic may be calculated, using (5.6), (5.7), and (5.8), and

$$E_0[T_L] = \sum_{k=1}^{K} k E_0[R_{k..}]$$

$$Var_0[T_L] = \sum_{k=1}^{K} k^2 Var_0[R_{k..}] + \sum_{i=1}^{K} \sum_{k=1, k \neq i}^{K} ik Cov_0[R_{i..}, R_{k..}].$$

In the balanced case, with $M_{kl} = M \; \forall k, l$, moments simplify to

$$E_0[T_L] = (K+1)KLM(KM+1)/4$$

$$Var_0[T_L] = LM^2(KM+1)\left(\sum_{k=1}^{K} k^2(K-1) - \sum_{i=1}^{K}\sum_{k=1,k\neq i}^{K} ik\right)$$

$$= LM^2(KM+1)\left(K\sum_{k=1}^{K} k^2 - (\sum_{i=1}^{K} i)^2\right)$$

$$= K^2(K+1)LM^2(KM+1)(K-1)/12,$$

using (3.20).

Example 5.6.1 *Refer again to the expense variability data of Example 5.4.1. Test for an ordered effect of income, treating categories as blocks. Rank sums are* $23, 36, 53, 57, 70, 70, 83$, *and the statistic is* $23 \times 1 + 36 \times 2 + 53 \times 3 + 57 \times 4 + 70 \times 5 + 70 \times 6 + 83 \times 7 = 1833$, *with expected value 1568 and variance under the null hypothesis 1829.333. The two-sided p-value is* 2.90×10^{-10}. *Page's test may be performed in R using*

```
library(crank); page.trend.test(expensesd[,-1],FALSE)
```

Page (1963) presents only the case with M_{kl} all 1, and provides a table of the distribution of T_L for small values of K and L. For larger values of K and L than appear in the table, Page (1963) suggests a Gaussian approximation.

The scores in (5.10) are equally spaced. This is often a reasonable choice in practice. When the M_{kl} are not all the same, imbalance among numbers of ranks summed may change the interpretation of these scores, and a preferred statistic definition to replace (5.10) is

$$T_L^* = \sum_{k=1}^{K} k \bar{R}_{k..}$$

with moments given by

$$\mathrm{E}\left[\bar{R}_{k..}\right] = \sum_{l=1}^{L} \left[M_{kl} \left(\sum_{j=1}^{K} M_{jl} + 1 \right) / 2 \right] / \sum_{l=1}^{L} M_{kl},$$

$$\mathrm{Var}\left[\bar{R}_{k..}\right] = \frac{\sum_{l=1}^{L}\{M_{kl}(\sum_{j \neq k} M_{jl})(\sum_{j=1}^{K} M_{jl} + 1)\}}{12(\sum_{l=1}^{L} M_{kl})^2},$$

and

$$\mathrm{Cov}\left[\bar{R}_{k..}, \bar{R}_{m..}\right] = -\sum_{l=1}^{L}[M_{kl}M_{ml}] \left(\sum_{j=1}^{K} M_{jl} + 1 \right) / \left(12 \sum_{l=1}^{L} M_{kl} \sum_{l=1}^{L} M_{ml} \right).$$

Example 5.6.2 *Refer again to the chicken weight gain data of Example 5.4.2. This data set is balanced, but one could ignore the balance and numerically invert the variance matrix of all rank sums except the last. In this case, test for an ordered effect of protein level on weight gain, blocking on protein source. Perform the test using*

```
library(MultNonParam)
attach(chicken)
cat('\n Page test with replicates  \n')
page.test.unbalanced(weight,lev,source)
detach(chicken)
```

> *In this balanced case, rank mean expectations are all 6.5, variances are 1.083, covariances are −0.542. The rank means by treatment level are 7.625, 7.250, 4.625, giving an overall statistic value of 36, compared with a null expectation of 39 and null variance of 3.25; the p-value is 0.096. Do not reject the null hypothesis.*

5.7 Exercises

1. The data set

    http://ftp.uni-bayreuth.de/math/statlib/datasets/schizo

    reflects an experiment using measurements used to detect schizophrenia, in both schizophrenic and non-schizophrenic patients. Various tests are given. Pick those data points with CS in the second column and compare the first and second gain rations, in the third and fourth columns, respectively. For this question, select only non-schizophrenic patients.

    a. Test the null hypothesis that first and second gain ratios (in the third and fourth columns, respectively) have the same median using the sign test.

    b. Test the null hypothesis that first and second gain ratios (in the third and fourth columns, respectively) have the same median using the signed rank test.

2. The data set

    http://ftp.uni-bayreuth.de/math/statlib/datasets/schizo

    contains data from an experiment using measurements used to detect schizophrenia, on non-schizophrenic patients. Various tests are given. Pick those data points with CS in the second column and compare the first and second gain ratios, in the third and fourth columns, by taking their difference. Test the null hypothesis of zero median difference using the Wilcoxon signed rank test, for the non-schizophrenic patients.

3. Calculate the asymptotic relative efficiency for the Wilcoxon signed rank statistic relative to the one-sample t-test (which you should approximate using the one-sample z-test). Do this for observations from

    a. the distribution uniform on the interval $[\theta - 1/2, \theta + 1/2]$,

    b. the logistic distribution, symmetric about θ, with variance $\pi^2/3$ and density $\exp(x - \theta)/(1 + \exp(x - \theta))^2$,

Exercises

c. and the standard normal distribution, with variance 1 and expectation θ.

4. The data set at

    http://stat.rutgers.edu/home/kolassa/Data/yarn.dat

    represents strengths of yarn of two types from six bobbins. This file has three fields: strength, bobbin, and type. Apply the balanced variant of Friedman's test to determine whether strength varies by type, controlling for bobbin.

5. The data at

    http://lib.stat.cmu.edu/datasets/CPS_85_Wages

    reflects wages from 1985. The first 42 lines of this file contain a description of the data set, and an explanation of variables; delete these lines first, or skip them when you read the file. All fields are numeric. The tenth field is sector, the fifth field is union membership, and the sixth field is hourly wage; you can skip everything else. Test for a difference in wage between union members and non-members, blocking on sector.

6

Bivariate Methods

Suppose that independent random vectors (X_i, Y_i) all have the same joint density $f_{X,Y}(x,y)$. This chapter investigates the relationship between X_i and Y_i for each i. In particular consider testing the null hypothesis $f_{X,Y}(x,y) = f_X(x)f_Y(y)$, without specifying an alternative hypothesis for $f_{X,Y}$, or even without knowledge of the null marginal densities.

This null hypothesis is most easily tested against the alternative hypothesis that, vaguely, large values of X are associated with large values of Y (or vice versa). Furthermore, if the null hypothesis is not true, the strength of the association between X_i and Y_i must be measured.

6.1 Parametric Approach

Before developing nonparametric approaches to assessing relationships between variables, this section reviews a standard parametric approach, that of the Pearson correlation (Edgeworth, 1893):

$$r_P = \frac{\sum_{j=1}^n (X_j - \bar{X})(Y_j - \bar{Y})}{\sqrt{\sum_{j=1}^n (X_j - \bar{X})^2 \sum_{j=1}^n (Y_j - \bar{Y})^2}}, \quad (6.1)$$

derived as what would now be termed the maximum likelihood estimator of the correlation for the bivariate Gaussian distribution. This gives the slope of the least squares line fitting Y to X, after scaling both variables by standard deviation. The Cauchy-Schwartz Theorem says that r_P is always in $[-1, 1]$. Perfect positive or negative linear association is reflected in a value for r_P of 1 or -1 respectively. Furthermore, there are a variety of early exact and approximate distributional results for r_P assuming that the observations are multivariate Gaussian.

6.2 Permutation Inference

Even when observations are not multivariate Gaussian, the measure (6.1) remains a plausible summary of association between variables. This section presents a distributional result for the Pearson correlation that does not require knowledge of the multivariate distribution of observations.

Under the null hypothesis that X_j is independent of Y_j, for all j, then every permutation of the Y values among experimental units, with the X values held fixed, is equally likely. Under this permutation distribution, conditional on the collections $\{X_1, \ldots, X_n\}$ and $\{Y_1, \ldots, Y_n\}$, the variance of r_P may be calculated directly. Note that r_P depends only on the differences between the observations and their means, and so, without loss of generality, assume that $\bar{X} = \bar{Y} = 0$. Let $(Z_1, \ldots, Z_n)$ be a random permutation of $\{Y_1, \ldots, Y_n\}$. Then $\mathrm{Var}\,[Z_1] = \sum_i Y_i^2/n$; denote this by σ_Y^2. Also

$$\mathrm{Cov}\,[Z_1, Z_2] = \sum_{i \neq j} \frac{1}{n(n-1)} Y_i Y_j$$

$$= \sum_{i,j} \frac{1}{n(n-1)} Y_i Y_j - \sum_i \frac{1}{n(n-1)} Y_i^2 = -\frac{1}{n-1} \sigma_Y^2.$$

So

$$\mathrm{Var}\left[\sum_j X_j Z_j\right] = \sum_j X_j^2 \mathrm{Var}\,[Z_j] + \sum_{j \neq i} X_i X_j \mathrm{Cov}\,[Z_j, Z_i]$$

$$= \sum_j X_j^2 \sigma_Y^2 - \frac{[\sum_{j,i} X_i X_j - \sum_i X_i^2]\sigma_Y^2}{n-1}$$

$$= \sum_j X_j^2 \sigma_Y^2 + \sum_i X_i^2 \sigma_Y^2/(n-1)$$

$$= n^2 \sigma_X^2 \sigma_Y^2/(n-1),$$

where $\sigma_X^2 = \sum_i \frac{1}{n} X_i^2$. So

$$\mathrm{Var}\,[r_P] = 1/(n-1), \qquad (6.2)$$

under the permutation distribution (Hotelling and Pabst, 1936). This value was determined by Student (1908), after fitting a parametric model to empirical data, rounding parameter estimates to values more consistent with intuition, and calculating the moments of this empirical distribution. Higher-order moments were determined by David et al. (1951) using a method of proof similar to that presented above for the variance.

This result suggests a test of the null hypothesis of independence versus the two-sided alternative at level α using r_P, that rejects the null hypothesis if

$$|r_P| > z_{\alpha/2}/\sqrt{n-1}. \qquad (6.3)$$

6.3 Nonparametric Correlation

The Pearson correlation p_r is designed to measure linear association between two variables, and fails to adequately reflect non-linear association. Furthermore, a family of distributional results, not recounted in this volume, depend on data summarized being multivariate Gaussian. Various nonparametric alternatives to the Pearson correlation have been developed to avoid these drawbacks.

6.3.1 Rank Correlation

Instead of calculating the correlation for the original variables, calculate the correlation of the ranks of the variables (Spearman, 1904). Under the null hypothesis, each X rank should be be equally likely to be associated with each Y rank. Under the alternative, extreme ranks should be associated with each other. Let R_j be the rank of the Y value associated with $X_{(j)}$. Define the Spearman Rank correlation as

$$r_S = \frac{\sum_{j=1}^n (j-(n+1)/2)(R_j-(n+1)/2)}{\sqrt{\left(\sum_{j=1}^n (j-(n+1)/2)^2\right)\left(\sum_{j=1}^n (R_j-(n+1)/2)^2\right)}};$$

that is, r_S is the Pearson correlation on ranks. Exact agreement for ordering of ranks results in a rank correlation of 1, exact agreement in the opposite direction results in a rank correlation of -1, and the Cauchy-Schwartz theorem indicates that these two vales are the extreme possible values for the rank correlation.

The sums of squares in the denominator have the same value for every data set, and the numerator can be simplified. Note that

$$\sum_{j=1}^n (j-(n+1)/2)^2 = \sum_{j=1}^n j^2 - n(n+1)^2/4$$
$$= n(n+1)(2n+1)/6 - n(n+1)^2/4$$
$$= n\left(n^2-1\right)/12.$$

Similarly, $\sum_{j=1}^n (R_j-(n+1)/2)^2 = n\left(n^2-1\right)/12$. Furthermore, $\sum_{j=1}^n (j-(n+1)/2)(n+1)/2 = 0$, and

$$r_S = \frac{12}{n(n^2-1)}\sum_{j=1}^n (j-(n+1)/2)R_j = \frac{12}{n(n^2-1)}\left(\sum_{j=1}^n jR_j - n(n+1)^2/4\right). \tag{6.4}$$

Hoeffding (1948) provides a central limit theorem for the permutation distribution for both r_P and r_S, including under alternative distributions. P-values may be approximated by dividing the observed correlations by $\sqrt{n-1}$, and comparing to a standard Gaussian distribution, but this approximation has poor relative behavior for small test levels. Best and Roberts (1975) correct the Gaussian approximation using an Edgeworth approximation (Kolassa, 2006); that is, they determine constants κ_1, κ_2, κ_3, and κ_4 such that

$$P\left[\frac{r_s - \kappa_1}{\sqrt{\kappa_2/n}} \leq r\right] \approx \Phi(r) - \phi(r)\left(\frac{\kappa_3 h_2(r)}{\kappa_2^{3/2} 6\sqrt{n}} + \frac{\kappa_3^2 h_5(r)}{72\kappa_2^3 n} + \frac{\kappa_4 h_3(r)}{24\kappa_2^2 n}\right), \quad (6.5)$$

with the maximum error in equation (6.5) bounded by a constant divided by $n^{3/2}$. Here $h_j(r)$ are known polynomials called Hermite polynomials, and constants κ_j are related to the moments of r_s. When κ_3 is calculated for a symmetric distribution, its value is zero, and approximation (6.5) effectively contains only its first and last terms. This is the case for many applications of (6.5) to rank-based statistics, including the application to r_s.

Example 6.3.1 *Consider again the twin brain data of Example 5.2.1, plotted in Figure 6.1 As before, the data set* brainpairs *has 10 records, reflecting the results from 10 pairs of twins, and is plotted in Figure 6.1 via*

```
attach(brainpairs); plot(v1,v2, xlab="Volume for Twin 1",
ylab="Volume for Twin 2",main="Twin Brain Volumes")
```

Ranks for twin brains are given in Table 6.1. The sum in the second factor of (6.4) is

$$1 \times 1 + 4 \times 2 + 9 \times 9 + 5 \times 5 + 3 \times 4 + 6 \times 7 + 7 \times 3 + 10 \times 10 + 2 \times 6 + 8 \times 8 = 366,$$

and so the entire second factor is $366 - 10 \times 11^2/4 = 63.5$, *and the Spearman correlation is* $(12/(10 \times 99)) \times 63.5 = 0.770$. *Observed correlations may be calculated in R using*

```
cat('\n Permutation test for twin brain data  \n')
attach(brainpairs)
obsd<-c(cor(v1,v2),cor(v1,v2,method="spearman"))
```

to obtain the Pearson correlation (6.1) 0.914 and the Spearman correlation (6.4) 0.770. Testing the null hypothesis of no association may be performed using a permutation test with either of these measures.

```
out<-array(NA,c(2,20001))
dimnames(out)[[1]]<-c("Pearson","Spearman")
for(j in seq(dim(out)[2])){
    newv1<-sample(v1)
    out[,j]<-c(cor(newv1,v2),cor(newv1,v2,method="spearman"))
}
```

```
cat("\n Monte Carlo One-Sided p value\n")
apply(apply(out,2,">=",obsd),1,"mean")
```

to obtain p-values 1.5×10^{-4} and 6.1×10^{-3}. The asymptotic critical value, from (6.3), is given by

```
cat("\n Asymptotic Critical Value\n")
-qnorm(0.025)/sqrt(length(v1)-1)
```

which gives 0.6533. Permutation tests based on either correlation method reject the null hypothesis.

```
c(cor.test(v1,v2)$p.value,
    cor.test(v1,v2,method="spearman")$p.value)
detach(brainpairs)
```

giving p-value approximations 2.1×10^{-4} and 1.37×10^{-2} respectively.

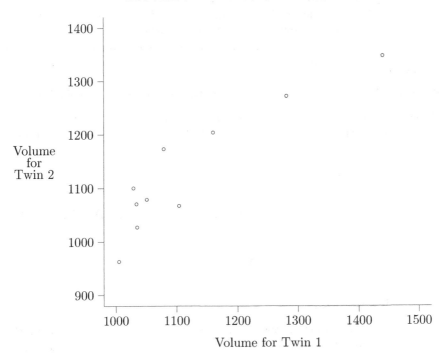

FIGURE 6.1: Twin Brain Volumes

TABLE 6.1: Twin brain volume ranks

Pair	1 2 3 4 5 5 7 8 9 10
Rank of First	1 4 9 5 3 6 7 10 2 8
Rank of Second	1 2 9 5 4 7 3 10 6 8

Note that

$$\sum_{j=1}^{n} jR_j = \sum_{j=1}^{n} j\left(1 + \sum_{i \neq j} I(Y_j > Y_i, X_j > X_i)\right)$$

$$= n(n+1)/2 + \sum_{j=1}^{n} j \sum_{i \neq j} I(Y_j > Y_i, X_j > X_i).$$

Some algebra shows that

$$r_S = 1 - \left(\frac{6}{n(n^2-1)}\right) S \text{ for } S = \sum_{j=1}^{n} (j - R_j)^2. \tag{6.6}$$

This relationship will be exploited to give probabilities for r_S in §6.4.1.3.

Pearson (1907) criticizes the Spearman correlation, in part on the grounds that, as it is designed to reflect association even when the association is nonlinear, it also loses the interpretation as a regression parameter, even when the underlying association is linear.

6.3.1.1 Alternative Expectation of the Spearman Correlation

Under the null hypothesis of independence between X_j and Y_j, for all j, $\mathrm{E}\left[r_S\right] = 0$, and the variance is given by (6.2). Under the alternative hypothesis,

$$\mathrm{E}\left[r_S\right] = \frac{12}{n(n^2-1)}(n(n+1)/2 + \sum_{j=1}^{n} j \sum_{i \neq j} p_1 - n(n+1)^2/4)$$

$$= \frac{12}{n(n^2-1)}((n-1)n(n+1)p_1/2 - n(n^2-1)/4) = 3(2p_1 - 1)$$

for

$$p_1 = \mathrm{P}\left[Y_j > Y_i, X_j > X_i\right]. \tag{6.7}$$

6.3.2 Kendall's τ

Pairs of bivariate observations for which the X and Y values are in the same order are called <u>concordant</u>; (6.7) refers to the probability that a pair is concordant. Pairs that are not concordant are called <u>discordant</u>. Kendall (1938) constructs a new measure based on counts of concordant and discordant pairs. Consider the population quantity $\tau = 2p_1 - 1$, for p_1 of (6.7),

Nonparametric Correlation

called Kendall's τ; to emphasize the parallelism between this measure and other measures of association between two random variables, it will also be referred to below as Kendall's correlation measure. This quantity reflects the probability of a concordant pair minus the probability of a discordant pair. Denote the number of concordant pairs by

$$U = \sum_{i<j} Z_{ij},$$

for $Z_{ij} = I((X_j - X_i)(Y_j - Y_i) > 0)$. Then $U^\dagger = n(n-1)/2 - U$ is the number of discordant pairs; this number equals the number of rearrangements necessary to make all pairs concordant. Estimate τ by the excess of concordant over discordant pairs, divided by the maximum:

$$r_\tau = \frac{U - (n(n-1)/2 - U)}{n(n-1)/2} = \frac{4U}{n(n-1)} - 1. \tag{6.8}$$

Note that $\mathrm{E}\,[U] = n(n-1)p_1/2$, for p_1 of (6.7), and $\mathrm{E}\,[r_\tau] = 2p_1 - 1$. The null value of p_1 is half, recovering

$$\mathrm{E}_0\,[U] = n(n-1)/4, \quad \mathrm{E}_0\,[r_\tau] = 0. \tag{6.9}$$

As with the Spearman correlation r_S, Kendall's correlation r_τ may be used to test the null hypothesis of independence, relying on its asymptotic Gaussian distribution, but the test requires the variance of r_τ. Note that

$$\mathrm{Var}\,[U] = \sum_{i<j} \mathrm{Var}\,[Z_{ij}] + \sum{}^{*} \mathrm{Cov}\,[Z_{ij}, Z_{kl}].$$

Here $\sum^*$ the sum over three distinct indices $i < j$, $k < l$. This sum consists of $n^2(n-1)^2/4 - n(n-1)/2 - n(n-1)(n-2)(n-3)/4 = n(n-1)(n-2)$ terms. Hence

$$\begin{aligned}\mathrm{Var}\,[U] &= \sum_{i<j}(p_1 - p_1^2) + \sum{}^{*}(p_3 - p_1^2) \\ &= n(n-1)(p_1 - p_1^2)/2 + n(n-1)(n-2)(p_3 - p_1^2),\end{aligned}$$

where

$$p_3 = \mathrm{P}\,[(X_1 - X_2)(Y_2 - Y_1) \geq 0, (X_1 - X_3)(Y_3 - Y_1) \geq 0].$$

The null value of p_3 is $\tfrac{5}{18}$, as can be seen by examining all 36 pairs of permutations of $\{1, 2, 3\}$. Hence

$$\mathrm{Var}_0\,[U] = n(n-1)/8 + n(n-1)(n-2)/36 = n(n-1)(5+2n)/72, \tag{6.10}$$

and

$$\mathrm{Var}_0\,[r_\tau] = 2(2n+5)/(9n(n-1)). \tag{6.11}$$

The result of Hoeffding (1948) also proves that r_τ is approximately Gaussian, including under alternative distributions. El Maache and Lepage (2003) discuss the multivariate distribution of both r_τ and r_S from collections of variables.

Example 6.3.2 *Consider again the twin brain data of Example 5.2.1, with ranks in Table 6.1. Discordant pairs are 2 − 5, 2 − 9, 4 − 7, 4 − 9, 5 − 7, 5 − 9, 6 − 7, and 7 − 9. Hence 8 of 45 pairs are discordant, and the remainder, 37, are concordant. Hence $r_\tau = 4 \times 37/90 - 1 = 0.644$, from (6.8). This may be calculated using*

```
cor(brainpairs$v1,brainpairs$v2,method="kendall")
```

in R. A test of the null hypothesis of no correlation may be performed by calculating $z = r_\tau/\sqrt{2(2n+5)/(9n(n-1))} = 0.644/\sqrt{2 \times 25/810} = 0.644/0.248 = 2.60$; reject the null hypothesis of no association for a two-sided test of level 0.05. *Function* Kendall *in package* Kendall *in R uses a modification of the method of Best and Roberts (1975).*

Figure 6.2 shows two artificial data sets with functional relationships between two variables. In the first relationship, the two variables are identical, while in the second, the second variable is related to the arc tangent of the first. Both variables relationships show perfect association. The first relationship represents perfect linear association, while the second reflects perfect nonlinear association. Hence the Spearman and Kendall association measures are 1 for both relationships; the Pearson correlation is 1 for the first relationship, and 0.937 for the second relationship.

FIGURE 6.2: Plot of Linear and Curved Relationships

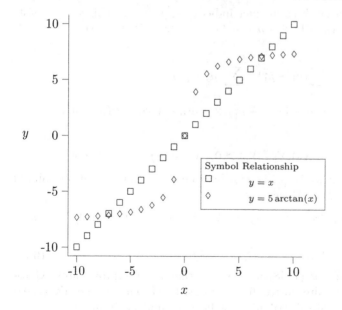

6.4 Bivariate Semi-Parametric Estimation via Correlation

Assume a linear model $Y_j = \beta_1 + \beta_2 X_i + e_i$, where e_i has some distribution with median 0 (making β_1 identifiable). Setting the correlation between $Y_j - \beta_2 X_i$ and X_i to zero, for any of the preceding correlation measures, may be used to estimate β_2. Furthermore, one might obtain a confidence interval by inverting test of the null hypothesis that the residuals are independent of the explanatory variables. This approach could be addressed using the Pearson, Spearman, or Kendall correlation. Using the Pearson approach gives the traditional least-squares estimate for β_2, and approximating the distribution of the errors as Gaussian leads to standard Gaussian theory inference.

6.4.1 Inversion of the Test of Zero Correlation

One might invert the test using correlation between X_i and $Y_i - \beta_2 X_i$, as a function of β_2, using any of the three preceding correlation methods, generically called r. That is, let $T(\beta_2) = r(\boldsymbol{X}, \boldsymbol{Y} - \beta_2 \boldsymbol{X})$, and note that this function is monotonic in β_2. The preceding paragraph discussed this approach for the Pearson correlation; now consider applying this approach to more general correlation measures. First, calculate t_L° and t_U°, such that $P\left[t_L^\circ < T(\beta_2, \text{data}) < t_U^\circ\right] < 1 - \alpha$, exactly as in (1.19). Then solve $T(\beta_2) = t_L^\circ$, $r(\beta_2) = t_U^\circ$.

When inverting correlation, inference on β_1 is impossible, because it washes out of ranks. One might estimate β_1 by setting the median of the residuals to zero, as is described below; some authors produce confidence intervals for β_1 via the bootstrap of Chapter 10.

6.4.1.1 Inversion of the Pearson Correlation

As noted above, the fully-parametric analysis, assuming bivariate Gaussian data, and using r_P, reduces to the standard inference from ordinary least squares regression. Inference under the permutation distribution avoids distributional assumptions on the data apart from independence.

Both determining critical values for r_P, and inverting the test statistic as a function of β_2 to give confidence bounds, are tricky for Pearson correlation, under the permutation distribution. Exact (non-Monte Carlo) probability calculations for the permutation distribution of the Pearson correlation are as difficult as enumerating all permutations. These calculations are often simplified by rounding data values to a lattice. Alternatively, since the null mean and variance are known, and since the distributions of these measures are approximately Gaussian, quantiles could be approximated using the Gaussian distribution. Also, r_P has steps of irregular size, and so translating $q_{\alpha/2}$ and $q_{1-\alpha/2}$ into values of β_2 is difficult.

6.4.1.2 Inversion of Kendall's τ

Calculations are easier for Kendall's τ. Best and Gipps (1974), using the algorithm of Kaarsemarker and van Wijngaarden (1953), present software for calculating all of the probability atoms, and hence quantiles, exactly for samples as large as 8, and with an Edgeworth approximation for larger samples. They note that, for $c(n, u)$ the count of permutations of $Y_1, \ldots, Y_n$ with u concordant pairs, then

$$c(n, u) = \sum_{s=u-n+1}^{u} c(n-1, s), \text{ for } n > 1, \ n(n-1)/2 \geq u \geq 0, \quad (6.12)$$

$$c(n, u) = 0 \text{ for } u < 0, \ c(0, 0) = 1,$$

$$c(n, u) = 0 \text{ for } u > n(n-1)/2. \quad (6.13)$$

Relation (6.12) is motivated by considering the possible effects of adding a final observation for a data set with $n - 1$ observations. Furthermore, $P[U = u] = c(n, u)/n!$. Derivation of the recursion is similar to that of (3.17).

A recursion for counts of permutations satisfying $U \leq u$ is identical, except that

$$c(n, u) = 1 \text{ for } u > n(n-1)/2$$

replaces (6.13).

Quantile $1 - \alpha$ of U can be calculated by inverting $P[U \leq u]$, for sample sizes corresponding to feasible exact calculations, or via

$$E_0[U] + z_\alpha \sqrt{\text{Var}_0[U]},$$

for moments given by (6.9) and (6.10). Jumps in r_τ, as a function of β_2, occur at pairwise slopes $W_{ij} = (Y_j - Y_i)/(X_j - X_i)$, since at these points the pairs $(X_i, Y_i - \beta_2 X_i)$ and $(X_j, Y_j - \beta_2 X_j)$ change from being concordant to discordant (Theil, 1950), and so an estimate $\hat{\beta}_2$ of β_2 is given by the median of these slopes, and the confidence interval endpoints are given as in §3.10.1. This last operation is more easily performed using the number of concordant pairs U, as in (6.8). Hence if t° satisfies $P_0[U \leq t^\circ] \geq \alpha/2$ and $P_0[U \geq t^\circ] \geq 1-\alpha/2$, and if $\tilde{W}_1, \ldots, \tilde{W}_{n(n-1)/2}$ are the ordered values of the pairwise slopes W_{ij}, then $(\tilde{W}_{t^\circ}, \tilde{W}_{n(n-1)/2+1-t^\circ})$ is a $1 - \alpha$ confidence interval for β_2. As noted above, the intercept β_1 may be estimated as that value $\hat{\beta}_1$ to give the residuals $Y_j - \hat{\beta}_1 - \hat{\beta}_2 X_j$ zero median; that is,

$$\hat{\beta}_1 = \text{smed}\left[Y_1 - \hat{\beta}_2 X_1, \ldots, Y_n - \hat{\beta}_2 X_n\right]. \quad (6.14)$$

Generally, $\hat{\beta}_1 \neq \text{smed}[Y_1, \ldots, Y_n] - \hat{\beta}_2 \text{smed}[X_1, \ldots, X_n]$. Sen (1968) investigates this procedure further.

This approach to estimation of β_2 only holds in contexts with no ties among the explanatory variable values.

Example 6.4.1 *Sen applies this procedure to the data set*

```
1  2  3  4 10 12 18
9 15 19 20 45 55 78
```

The following commands calculate a confidence interval for the slope parameter:

```
tt<-c(1,2,3,4,10,12,18); xx<-c(9,15,19,20,45,55,78)
out<-rep(NA,length(tt)*(length(tt)-1)/2)
count<-0
for(ii in seq(length(tt)-1)) for(jj in (ii+1):length(tt)){
    count<-count+1
    out[count]<-(xx[jj]-xx[ii])/(tt[jj]-tt[ii])
}
```

There are $7 \times 6/2 = 21$ pairwise slopes $\tilde{W}_j$:

```
1.00 2.50 3.67 3.71 3.75 3.83 3.93 3.94 4.00 4.00
4.00 4.00 4.06 4.12 4.14 4.17 4.18 4.38 5.00 5.00 6.00
```

The estimate of β_2 in the regression model med$[X_j] = \beta_1 + \beta_2 t_j$ is the median of these pairwise slopes, 4. Confidence intervals may be exhibited using quantiles of the number of concordant pairs:

```
library(MultNonParam); qconcordant(0.025,7)
```

giving 4 as the quantile. Hence the 0.95 confidence interval is $(\tilde{W}_4, \tilde{W}_{18}) = (3.71, 4.38)$. These calculations may be performed using `theil(xx,tt)`. *It may also be performed using* `theilsen(xx~tt)` *from the package* `deming`, *which also gives a confidence interval for the intercept term. The intercept is estimated by (6.14). Plotting may be done via*

```
library(deming); plot(tt,xx);abline(theilsen(xx~tt))
```

Results are in Figure 6.3.

6.4.1.3 Inversion of the Spearman Correlation

Calculation for the Spearman correlation r_S is more complicated. van de Wiel and Di Bucchianico (2001) provide tools for the exact calculation of the null sampling distribution of the Spearman correlation r_S; this algorithm is implemented in the R package `pspearman`, which gives probabilities for the random variable S of (6.6). Closed-form inversion of the function $T(\beta_2) = r_S(\boldsymbol{X}, \boldsymbol{Y} - \beta_2 \boldsymbol{X})$ is more difficult.

FIGURE 6.3: Theil-Sen Estimator for Artificial Example

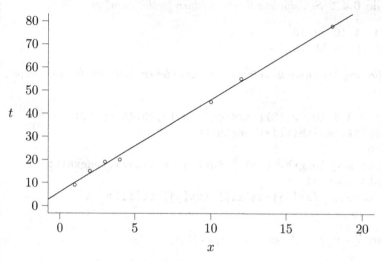

Example 6.4.2 *Consider data on changes in systolic and diastolic blood pressure after treatment by an angiotensin-converting enzyme inhibitor (Cox and Snell, 1981, Example E), given at*

http://www.stat.ucla.edu/ rgould/datasets/bloodpressure.dat .

We calculate the estimates and confidence intervals for the linear dependence of change in diastolic blood pressure, Y_i, on the change in systolic blood pressure, X_i. Pearson, Spearman, and Kendall correlations between change in systolic blood pressure, and the change in diastolic blood pressure, are 0.105, 0.144, and 0.067 respectively. The three correlations between X_i and $Y_i - \beta_2 X_i$, as a function of the slope, are plotted using

```
bp<-as.data.frame(scan('bloodpressure.dat',skip=1,
  what=list(spb=0,spa=0,spd=0,dpb=0,dpa=0,dpd=0)))
library(NonparametricHeuristic)
ci<-invertcor(bp$spd,bp$dpd)
```

to generate Figure 6.4. The relationship for the Pearson correlation is smooth; the relationships for the Spearman and Kendall correlations are step functions. Horizontal lines corresponding to asymptotic Gaussian critical values for the test of zero correlation are plotted; the permutation variance (6.2) is used for the Pearson and Spearman correlation. The two lines are slightly offset in the picture to display both lines. The Kendall correlation uses variance (6.11). Vertical lines pass through the point at

Bivariate Semi-Parametric Estimation via Correlation

which the horizontal lines cross the correlation curve, and their intersections with the horizontal axis determine the end points of the regression confidence interval. Recall that the Theil estimator is the inversion of the Kendall correlation. Figure 6.5 displays the results of

```
attach(bp)
plot(spd,dpd,main="Blood Pressure Change", ylab="Diastolic",
   xlab="Systolic")
library(deming)#For theilsen
tsout<-theilsen(dpd~spd)
abline(tsout); abline(lm(dpd~spd),lty=2)
legend(median(spd),median(dpd),lty=1:2,
   legend=c("Inversion of tau","Least squares"))
detach(bp)
```

producing regression lines.

FIGURE 6.4: Construction of Regression Estimates for Blood Pressure Data

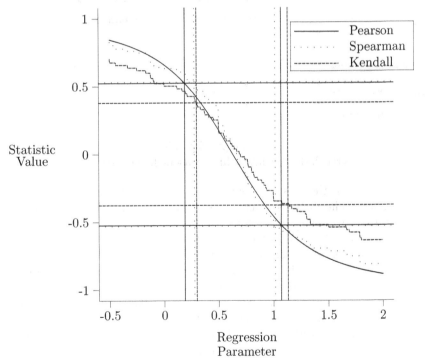

Statistic Quantiles are approximated using Normal

FIGURE 6.5: Blood Pressure Data

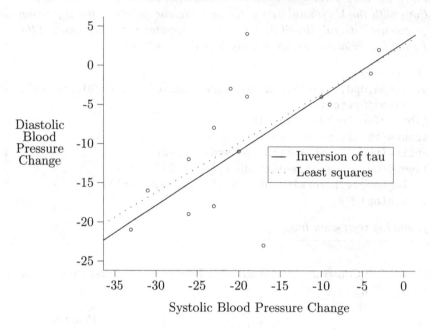

6.5 Exercises

1. The data set

    http://ftp.uni-bayreuth.de/math/statlib/datasets/federalistpapers.txt

    represents the result from an analysis of a series of documents. The first column gives document number, the second gives the name of a text file, the third gives a group to which the text is assigned, the fourth represents a measure of the use of first person in the text, the fifth presents a measure of inner thinking, the sixth presents a measure of positivity, and the seventh presents a measure of negativity. There are other columns that you can ignore. (The version at Statlib, above, has odd line breaks. A reformatted version can be found at stat.rutgers.edu/home/kolassa/Data/federalistpapers.txt). Plot use of positivity versus negativity. Calculate the three correlation measures, and relate their relative values to the shape of the curve.

    a. Test at $\alpha = .05$ the null hypothesis of zero population Spearman correlation for these two variables, versus the general alternative.

Exercises 127

   b. Give an estimate and a 95% confidence interval for the slope of
   Positivity on Negativity.

2. The data set

   http://stat.rutgers.edu/home/kolassa/Data/twinbrain.dat

   reflects brain volumes of twins. Using the method of Theil, estimate
   the slope parameter in the regression of second brain volume on the
   first brain volume, and give a confidence interval.

3. The data set

   http://ftp.uni-bayreuth.de/math/statlib/datasets/schizo

   reflects an an experiment using measurements used to detect
   schizophrenia, in both schizophrenic and non-schizophrenic patients.
   Results of various eye-tracking tests are given. The second column
   gives eye tracking target type. For all parts of this question select
   subjects with target type CS and TR. Gain ratios are given in various columns; for the balance of this question, use the first of these,
   found in the third column. Remove any subjects without both CS
   and TR measurements.

   a. Calculate the Spearman and Kendall correlations for CS and TR
   measurements.

   b. Give a 95% confidence interval for the slope of the relation of CS
   on TR, using the Thiel method.

7
Multivariate Analysis

Suppose that one observes n subjects, indexed by $i \in \{1, \ldots, n\}$, and, for subject i, observes responses X_{ij}, indexed by $j \in \{1, \ldots, J\}$. Potentially, covariates are also observed for these subjects.

This chapter explores explaining the multivariate distribution of X_{ij} in terms of these covariates. Most simply, these covariates often indicate group membership.

7.1 Standard Parametric Approaches

When data vectors may be treated as approximately multivariate Gaussian, the following standard techniques may be applied.

7.1.1 Multivariate Estimation

Often one wants to learn about a the vector of population central values for each of the j responses on the various subjects. In this section, assume that the vectors are independent and identically distributed.

Standard parametric analyses presuppose distributions of data well-enough behaved that location can be well-estimated using a sample mean. Denote the mean by the vector $\bar{X}$, where component j of this vector is given by $\bar{X}_j = \sum_{i=1}^{n} X_{ij}/n$. Then $\bar{X}$ is the method of moments estimator for $\boldsymbol{\mu} = \mathrm{E}\left[\boldsymbol{X}_i\right]$. Assumptions guaranteeing that $\bar{X}$ has an asymptotically Gaussian distribution generally include the existence of some moment of the distribution greater than the second moment.

7.1.2 One-Sample Testing

In this section, consider testing the null hypothesis that the vector $\boldsymbol{\mu}$ of expectations takes on some value specified in advance; without loss of generality, take this value to be $\mathbf{0}$. Still assuming that the observations have a multivariate Gaussian distribution, then $\bar{X}$ is approximately multivariate Gaussian. First consider the case in which one knows the variance matrix $\boldsymbol{\Sigma} = \mathrm{Var}\left[(X_{i1}, \ldots, X_{iJ})\right]$, and assume that $\boldsymbol{\Sigma}$ is nonsingular. Then one can

use as a test statistic
$$T^2 = \bar{\boldsymbol{X}}^\top (\boldsymbol{\Sigma}/n)^{-1} \bar{\boldsymbol{X}}, \tag{7.1}$$
and its distribution under the null hypothesis is χ_J^2.

Dropping the multivariate Gaussian assumption, if $\boldsymbol{\Sigma}$ is unknown, and if one can estimate it as $\hat{\boldsymbol{\Sigma}}$ using the usual sum of squares, then
$$T^2 = \bar{\boldsymbol{X}}^\top (\hat{\boldsymbol{\Sigma}}/n)^{-1} \bar{\boldsymbol{X}} \tag{7.2}$$
has an F distribution, with J numerator degrees of freedom (Hotelling, 1931). If the distribution of $(X_{i1}, \ldots, X_{iJ})$ has a density and a nonsingular variance matrix, then $\mathrm{P}\left[\hat{\boldsymbol{\Sigma}}\text{ singular}\right] = 0$. If $\boldsymbol{\Sigma}$ unknown, and is best estimated by a nonsingular $\tilde{\boldsymbol{\Sigma}}$, which is other than the sum of squares estimator, then generally (7.2) is approximately χ_J^2. These techniques require that $\bar{\boldsymbol{X}}$ is approximate multivariate Gaussian. This assumption is stronger than the assumption that each margin is univariate Gaussian; a simulated example is given in Figure 7.1.

7.1.3 Two-Sample Testing

Suppose that observations may be divided into two groups of sizes M_1 and M_2, with the group for observation i indicated by $g_i \in \{1, 2\}$. Test the null hypothesis that the multivariate distributions in the two groups are identical; note that this implies identical variance matrices. Let $\bar{\boldsymbol{X}}_k$ be the vector of sample means for observations in group k, with components $\bar{X}_{kj} = \sum_{i|g_i=k} X_{ij}/M_k$. Let $\hat{\Sigma}_{k,j,j'}$ be the sample covariance for the group k values between responses j and j': $\hat{\Sigma}_{k,j,j'} = \sum_{i|g_i=k}(X_{ij} - \bar{X}_k)(X_{ij'} - \bar{X}'_j)/(M_k - 1)$. Let $\hat{\Sigma}_{j,j'}$ be the pooled sample covariance for the all observations: $\hat{\Sigma}_{j,j'} = ((M_1 - 1)\hat{\Sigma}_{1,j,j'} + (M_2 - 1)\hat{\Sigma}_{2,j,j'})/(M_1 + M_2 - 2)$. Then the Hotelling two-sample statistic
$$T^2 = (\bar{\boldsymbol{X}}_1 - \bar{\boldsymbol{X}}_2)^\top ((1/M_1 + 1/M_2)\hat{\boldsymbol{\Sigma}})^{-1}(\bar{\boldsymbol{X}}_2 - \bar{\boldsymbol{X}}_2) \tag{7.3}$$
measures the difference between sample mean vectors, in a way that accounts for sample variance, and combines the response variables. Furthermore, under the null hypothesis of equality of distribution, and assuming that this distribution is multivariate Gaussian,
$$\frac{M_1 + M_2 - J - 1}{(M_1 + M_2 - 2)J} T^2 \sim \mathcal{F}_{J, M_1 + M_2 - J - 1}.$$

7.2 Nonparametric Multivariate Estimation

In the absence of such parametric assumptions, one might instead measure location using the <u>multivariate median</u>.

Nonparametric Multivariate Estimation

FIGURE 7.1: Univariate Normal Data That are Not Bivariate Normal

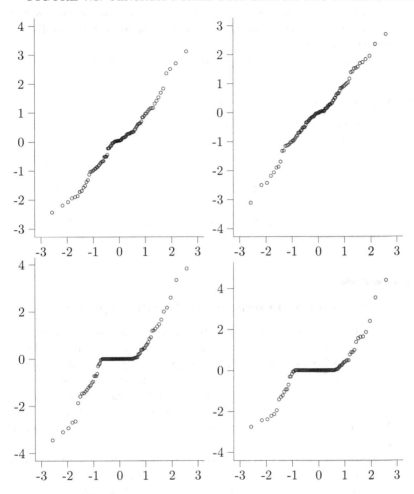

Define the multivariate population median $\boldsymbol{\nu}$ to be the vector of univariate medians, as defined in §2.3.1. An estimator smed $[\boldsymbol{X}_1, \ldots, \boldsymbol{X}_n]$ of the population multivariate median may be constructed as the vector of whose components are the separate marginal sample medians; that is, smed $[\boldsymbol{X}_1, \ldots, \boldsymbol{X}_n]$ = (smed $[X_{11}, \ldots, X_{1n}], \ldots,$ smed $[X_{J1}, \ldots, X_{Jn}]$).

Alternatively, one might define smed $[\boldsymbol{X}_1, \ldots, \boldsymbol{X}_n]$ so to minimize the sum of distances from the median:

$$\text{smed}\,[\boldsymbol{X}_1, \ldots, \boldsymbol{X}_n] = \operatorname*{argmin}_{\boldsymbol{\eta}} \sum_{i=1}^{n} \sum_{j=1}^{J} |X_{ij} - \eta_j|; \qquad (7.4)$$

that is, the estimate minimizes the sum of distances from data vectors to the parameter vector, with distance measured by the sum of absolute values of component-wise differences. Because one can interchange the order of summation in (7.4), the minimizer in (7.4) is the vector of component-wise minimizers. Furthermore, the minimizer for each component is the traditional univariate median as above.

A summary of multivariate median concepts is given by Small (1990).

7.2.1 Equivariance Properties

In the univariate case (that is, $J = 1$), both the mean and the median are equivariant with respect to affine transformations of the raw data, as seen in §2.1.1 and §2.3.1. Equivariance to affine transformations in the multivariate case holds for the mean: for a vector a and a matrix B with J columns, and for $Y_j = a + BX_j$ for all j, then $\bar{Y} = a + B\bar{X}$. A similar equality fails to hold for smed $[X_1, \ldots, X_n]$ and smed $[Y_1, \ldots, Y_n]$, unless B is diagonal; hence the multivariate median is not equivariant under affine transformations.

7.3 Nonparametric One-Sample Testing Approaches

Consider a null hypothesis stating that the marginal median vector ν takes on a value specified in advance; without loss of generality, take this to be zero. In the multivariate Gaussian context, the statistic (7.2) represents the combination of separate location test statistics for the various components of the random vectors, and its distribution depends on multivariate normality of the underlying observations; an analogous statistic combining the various dimensions of X_i that does not depend on parametric assumptions is constructed in this section.

A nonparametric hypothesis test can be constructed by assembling component-wise nonparametric statistics into a vector T, analogous to $\bar{X}$, and centered so that $E_0[T] = 0$. One might combine sign test statistics, or signed-rank statistics if one assumes symmetry, often in the context of paired data. That is,

$$T_j(X_{1j}, \ldots, X_{nj}) = \sum_{i=1}^n \tilde{s}(X_{ij} > 0), \qquad (7.5)$$

for $\tilde{s}(u) = \begin{cases} 1 & \text{for } u > 0 \\ -1 & \text{f } u \not< 0 \end{cases}$. Or, define $R_{ij}(X)$ to be the marginal rank of $|X_{ij}|$ among $\{|X_{1j}|, \ldots, |X_{nj}|\}$, and set

$$T_j(X_{1j}, \ldots, X_{nj}) = \sum_{i=1}^n R_{ij}(X)\tilde{s}(X_{ij} > 0). \qquad (7.6)$$

Nonparametric One-Sample Testing Approaches

A multivariate test statistic is constructed as a vector of univariate statistics,

$$T(X) = (T_1(X_{11}, \ldots, X_{n1}), \ldots, T_J(X_{1J}, \ldots, X_{nJ}))^\top.$$

Then combine components of T from (7.5) to give the multivariate sign test statistic, or from (7.6) to give the multivariate sign rank test. In either case, components are combined using

$$W = T^\top \Upsilon^{-1} T \qquad (7.7)$$

for $\Upsilon = \text{Var}_0[T]$. As in §2.3, in the case that the null location value is $\mathbf{0}$, the null distribution for the multivariate test statistic is generated by assigning equal probabilities to all 2^n modifications of the data set by multiplying the rows $(X_{i1}, \ldots, X_{iJ})$ by $+1$ or -1. That is, the null hypothesis distribution of $T(X)$ is generated by placing probability 2^{-n} on all of the 2^n elements of

$$\mathcal{X} = \{\tilde{X} \text{ an } n \times J \text{ matrix} | (\tilde{X}_{i1}, \ldots, \tilde{X}_{iJ}) = \pm(X_{i1}, \ldots, X_{iJ})\}. \qquad (7.8)$$

Test statistics (7.1) and (7.2) arose as quadratic forms of independent and identically distributed random vectors, and the variances included in their definitions were scaled accordingly. Statistic (7.3) is built using a more complicated variance; this pattern will repeat with nonparametric analogies to parametric tests.

Combining univariate tests into a quadratic form raises two difficulties. In previous applications of rank statistics, that is, in the case of univariate sign and signed-rank one-sample tests, in the case of two-sample Mann-Whitney-Wilcoxon tests, and in the case of of Kruskal-Wallis testing, all dependence of the permutation distribution on the original data was removed through ranking. This is not the case for T, since this distribution involves correlations between ranks of the various response vectors. These correlations are not specified by the null hypothesis. The separate tests are generally dependent, and dependence structure depends on distribution of raw observations. The asymptotic distribution of (7.7) relies on this dependence via the correlations between components of T. The correlations must be estimated.

Furthermore, the distribution of W of (7.7) under the null hypothesis is dependent on the coordinate system for the variables, but, intuitively, this dependence on the coordinate system might be undesirable. For example, suppose that (X_{1i}, X_{2i}) has an approximate multivariate Gaussian distribution, with expectation $\boldsymbol{\mu}$, and variance $\boldsymbol{\Sigma}$, with $\boldsymbol{\Sigma}$ known. Consider the null hypothesis $H_0 : \boldsymbol{\mu} = \mathbf{0}$. Then the canonical test is (7.1), and it is unchanged if the test is based on (U_i, V_i) for $U_i = X_{1i} + X_{2i}$ and $V_i = X_{1i} - X_{2i}$, with $\boldsymbol{\Sigma}$ modified accordingly. Hence the parametric analysis is independent of the coordinate system.

The first of this difficulty is readily addressed. Under H_0, the marginal sign test statistic (7.5) satisfies $T_j/\sqrt{n} \approx \Phi(0, 1)$. Conditional on the relative ranks of the absolute values of the observations, the permutation distribution is entirely specified, and conditional joint moments are calculated. Under the permutation distribution,

$$\hat{\sigma}_{jj'} = \sum_i \tilde{s}(X_{ij})\tilde{s}(X_{ij'})/n, \qquad (7.9)$$

and so the variance estimate used in (7.7) has components $\hat{v}_{jj'} = \hat{\sigma}_{jj'}/n$. Here again, the covariance is defined under the distribution that consists of 2^n random reassignment of signs to the data vectors, each equally weighted. As before, variances do not depend on data values, but covariances do depend on data values. The solution for the Wilcoxon signed-rank test is also available (Bennett, 1965).

Using the data to redefine the coordinate system may be used to address the second problem (Randles, 1989; Oja and Randles, 2004).

Combine the components of $T(X)$ to construct the statistic

$$W = W(X) = T^\top \hat{\Sigma}^{-1} T, \qquad (7.10)$$

using an estimate $\hat{\Sigma}$ of $\Sigma = \text{Var}[T]$ as in (7.9), or similarly for the signed-rank statistic.

The multivariate central limit theorem of Hájek (1960), and the quality of approximation to Υ, justifies approximating the null distribution of W by χ^2_J distribution. The test rejects the null hypothesis of zero component-wise medians when

$$W(X) > G_J^{-1}(1 - \alpha, 0)$$

for $G_J^{-1}(1 - \alpha, 0)$ the $1 - \alpha$ quantile of the χ^2_J distribution, with non-centrality parameter 0. Bickel (1965) discusses these (and other tests) in generality.

Example 7.3.1 *Consider the data of Example 6.4.2. We test the null hypothesis that the joint distribution of systolic and diastolic blood pressure changes is symmetric about $(0,0)$, using Hotelling's T^2 and the two asymptotic tests that substitutes signs and signed-ranks for data. This test is performed in R using*

```
# For Hotelling and multivariate rank tests resp:
library(Hotelling); library(ICSNP)
cat('\n One-sample Hotelling Test\n')
HotellingsT2(bp[,c("spd","dpd")])
cat('\n Multivariate Sign Test\n')
rank.ctest(bp[,c("spd","dpd")],scores="sign")
cat('\n Multivariate Signed Rank Test\n')
rank.ctest(bp[,c("spd","dpd")])
```

Nonparametric One-Sample Testing Approaches

P-values for Hotelling's T^2, the marginal sign rank test, and marginal sign test, are 9.839×10^{-6}, 2.973×10^{-3}, and 5.531×10^{-4}.

Tables 7.1 and 7.2 contain attained levels and powers for one-sample multivariate tests with two manifest variables of nominal level 0.05, from various distributions.

TABLE 7.1: Level of multivariate tests

Test	Sample Size 20			Sample Size 40		
	Normal	Cauchy	Laplace	Normal	Cauchy	Laplace
T2	0.04845	0.01550	0.04335	0.05025	0.01665	0.04605
Sign test	0.04215	0.04130	0.04365	0.04865	0.05055	0.04650
Sign rank test	0.04115	0.03895	0.04170	0.05065	0.04980	0.04745

TABLE 7.2: Power of multivariate tests

Test	Sample Size 20			Sample Size 40		
	Normal	Cauchy	Laplace	Normal	Cauchy	Laplace
T2	0.74375	0.08090	0.48460	0.97760	0.08845	0.79215
Sign test	0.70330	0.26290	0.55560	0.96955	0.52135	0.88700
Sign rank test	0.52615	0.32605	0.55025	0.87520	0.64720	0.89445

Tests compared are Hotelling's T^2 tests, and test (7.10) applied to the sign and signed-rank tests. Tests have close to their nominal levels, except for Hotelling's test with the Cauchy distribution; furthermore, the agreement is closer for sample size 40 than for sample size 20. Furthermore, the sign test power is close to that of Hotelling's test for Gaussian variables, and the signed-rank test has attenuated power. Both nonparametric tests have good power for the Cauchy distribution, although Hotelling's test performs poorly, and both perform better than Hotelling's test for Laplace variables.

Some rare data sets simulated to create Tables 7.1 and 7.2 include some for which Υ is estimated as singular. Care must be taken to avoid difficulties; in such cases, p-values are set to 1.

7.3.1 More General Permutation Solutions

One might address this problem using permutation testing. First, select an existing parametric test statistic $U(\boldsymbol{X})$, perhaps a Hotelling statistic, or a rank-based statistic. Under the permutation null distribution, the sampling distribution puts equal weight 2^{-n} to all 2^n values of the statistic evaluated at each element of (7.8); these 2^n values need not all be unique. For n large enough to make exhaustive evaluation prohibitive, a random subset of elements of (7.8) may be selected. The p-value is reported as the proportion of data sets with permuted signs having the test statistic value as large as, or larger than, that observed. In this way, the analysis of the previous subsection for the sign test, and by extension the signed-rank test, can be extended to general rank tests, including tests with data as scores.

7.4 Confidence Regions for a Vector Shift Parameter

Proceed analogously to the one-dimensional confidence interval construction as in §2.3.3. Introduce a shift parameter to move the data to an arbitrary point null hypothesis. In this one-sample case, one may apply the one sample test for the null hypothesis that marginal medians are $\mathbf{0}$ to the shifted data $\mathbf{X} - \mathbf{1}_n \otimes \boldsymbol{\mu}$, where $\mathbf{1}_n$ is a vector of ones of length n, and $\otimes$ is the outer product, so $\mathbf{1}_n \otimes \boldsymbol{\mu}$ is the matrix with entry μ_j in column j for all rows. That is, calculate $\mathbf{T}(\mathbf{X} - \mathbf{1}_n \otimes \boldsymbol{\mu})$ from the data set $\mathbf{X} - \mathbf{1}_n \otimes \boldsymbol{\mu}$. Calculate the variance matrix ((7.9) for the multivariate sign test) from the shifted data set. Calculate $W(\mathbf{X} - \mathbf{1}_n \otimes \boldsymbol{\mu})$ from (7.10). Then the random set

$$\mathcal{I} = \{\boldsymbol{\mu} | W(\mathbf{X} - \mathbf{1}_n \otimes \boldsymbol{\mu}) \leq \chi^2_{J,\alpha}\},$$

satisfies $P[\boldsymbol{\mu} \in \mathcal{I}] = 1 - \alpha$ using the test inversion argument of (1.16). This is the one-sample case of the region proposed by Kolassa and Seifu (2013).

Example 7.4.1 *Recall again the blood pressure data set of Example 6.4.2. Figure 7.2 is generated by*

`library(MultNonParam); shiftcr(bp[,c("dpd","spd")])`

and exhibits the 0.05 contour of p-values for the multivariate test constructed from sign rank tests for each of systolic and diastolic blood pressure, and forms a 95% confidence region. Note the lack of convexity.

7.5 Two-Sample Methods

Two-sample methods are generally of more interest than the preceding one-sample methods. Consider a multivariate data set X_{ij} for $i \in \{1, \ldots, M_1 + M_2\}$ and $j \in \{1, \ldots, J\}$, with data from the first sample occupying the first M_1 rows of this matrix, and data from the second sample occupying the last M_2 rows. Assume that the vectors $(X_{i1}, \ldots, X_{iJ})$ and $(X_{i'1}, \ldots, X_{i'J})$, for all i, are independent if $i \neq i'$. Assume further than the vectors $(X_{i1}, \ldots, X_{iJ})$ all

FIGURE 7.2: Median Blood Pressure Change Confidence Region

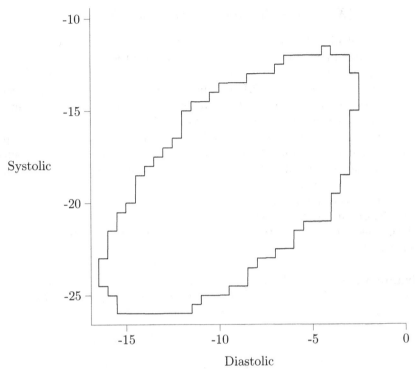

95% Confidence from Inversion of Bivariate Sign Rank Test

have the same distribution for $i \leq M_1$, and that the vectors $(X_{i1}, \ldots, X_{iJ})$ all have the same distribution for $i > M_1$. Let $\boldsymbol{g}$ be a vector indicating group membership; $g_i = 1$ if $i \leq M_1$, and $g_i = 2$ if $i > M_1$. As in §7.1, consider testing and confidence interval questions.

7.5.1 Hypothesis Testing

Combine the techniques for one-dimensional two-sample testing of Chapter 3 and for multi-dimensional one-sample testing of §7.1. Consider the null hypothesis H_0 that the distribution of $(X_{i1}, \ldots, X_{iJ})$ is the same for all i.

7.5.1.1 Permutation Testing

Under the null hypothesis of equality of distributions across the two groups, all assignments of the observed vectors among the two groups that keep the sizes of groups 1 and 2 at M_1 and M_2 respectively, are equally likely. Hence a permutation test may be constructed by evaluating the Hotelling statistic, or any other parametric statistic, at each of the $\binom{M_1+M_2}{M_1}$ such reassignments of the observations to the groups,

$$\mathcal{X}^* = \{(\boldsymbol{X}, \boldsymbol{g}) | \boldsymbol{g} \text{ has } M_1 \text{ entries that are 1 and } M_2 \text{ that are 2}\};$$

p-values are calculated by counting the number of such statistics with values equal to or exceeding the observed value, and dividing the count by $\binom{M_1+M_2}{M_1}$.

Other marginal statistics may be combined; for example, one might use the Max t-statistic, defined by first calculating univariate t-statistics for each manifest variable, and reporting the maximum. This statistic is inherently one-sided, in that it presumes an alternative in which each marginal distribution for the second group is systematically larger than that of the first group. Alternatively, one might take the absolute value of the t-statistics before maximizing. One might do a parallel analysis with either the maximum of Wilcoxon rank-sum statistics or the maximum of their absolute values, after shifting to have a null expectation zero.

Finally, one might apply permutation testing to the statistic (7.3), calculated on ranks instead of data values, to make the statistic less sensitive to extreme values.

Example 7.5.1 *Consider the data on wheat yields, in metric tons per hectare (Cox and Snell, 1981, Set 5), reflecting yields of six varieties of wheat grown at ten different experimental stations, from*

http://stat.rutgers.edu/home/kolassa/Data/set5.data .

Two of these varieties, Huntsman and Atou, are present at all ten stations, and so the analysis will include only these. Stations are included from three geographical regions of England; compare those in the north to those elsewhere. The standard Hotelling two-sample test may be performed in R using

```
wheat<-as.data.frame(scan("set5.data",what=list(variety="",
   y0=0,y1=0,y2=0,y3=0,y4=0,y5=0,y6=0,y7=0,y8=0,y9=0),
   na.strings="-"))
# Observations are represented by columns rather than by
# rows.  Swap this.  New column names are in first column.
dimnames(wheat)[[1]]<-wheat[,1]
wheat<-as.data.frame(t(wheat[,-1]))
dimnames(wheat)[[1]]<-c("E1","E2","N3","N4","N5","N6","W7",
   "E8","E9","N10")
wheat$region<-factor(c("North","Other")[1+(
   substring(dimnames(wheat)[[1]],1,1)!="N")],
   c("Other","North"))
attach(wheat)
plot(Huntsman,Atou,pch=(region=="North")+1,
   main="Wheat Yields")
```

Two-Sample Methods 139

```
legend(6,5,legend=c("Other","North"),pch=1:2)
```

Data are plotted in Figure 7.3. The normal-theory p-value for testing equality of the bivariate yield distributions in the two regions is given by

```
library(Hotelling)#for hotelling.test
print(hotelling.test(Huntsman+Atou~region))
```

The results of `hotelling.test` *must be explicitly printed, because the function codes the results as invisible, and so results won't be printed otherwise. The p-value is 0.0327. Comparing this to the univariate results*

```
t.test(Huntsman~region);t.test(Atou~region)
```

gives two substantially smaller p-values; in this case, treatment as a multivariate distribution did not improve statistical power. On the other hand, the normal quantile plot for Atou yields shows some lack of normality. Outliers do not appear to be present in these data, but if they were, they could be addressed by performing the analysis on ranks, either using asymptotic normality:

```
cat('Wheat rank test, normal theory p-values')
print(hotelling.test(rank(Huntsman)+rank(Atou)~region))
```

or using permutation testing to avoid the assumption of multivariate normality:

```
#Brute-force way to get estimate of permutation p-value for
#both T2 and the max t statistic.
cat('Permutation Tests for Wheat Data, Brute Force')
obsh<-hotelling.test(Huntsman+Atou~region)$stats$statistic
obst<-max(c(t.test(Huntsman~region)$statistic,
   t.test(Atou~region)$statistic))
out<-array(NA,c(1000,2))
dimnames(out)<-list(NULL,c("Hotelling","t test"))
for(j in seq(dim(out)[[1]])){
   newr<-sample(region,length(region))
   hto<-hotelling.test(Huntsman+Atou~newr)
   out[j,1]<-hto$stats$statistic>=obsh
   out[j,2]<-max(t.test(Huntsman~newr)$statistic,
      t.test(Atou~newr)$statistic)>=obst
}
apply(out,2,mean)
```

giving permutation p-values for the Hotelling and max-t statistics of 0.023 and 0.003 respectively. The smaller max-t statistic reflects the strong association between variety yields across stations. If one wants only the Hotelling statistic significance via permutation, one could use

```
print(hotelling.test(Huntsman+Atou~region,perm=T,
   progBar=FALSE))
```

The argument `progBar` will print a progress bar, if desired, and an additional argument controls the number of random permutations.

FIGURE 7.3: Wheat Yields

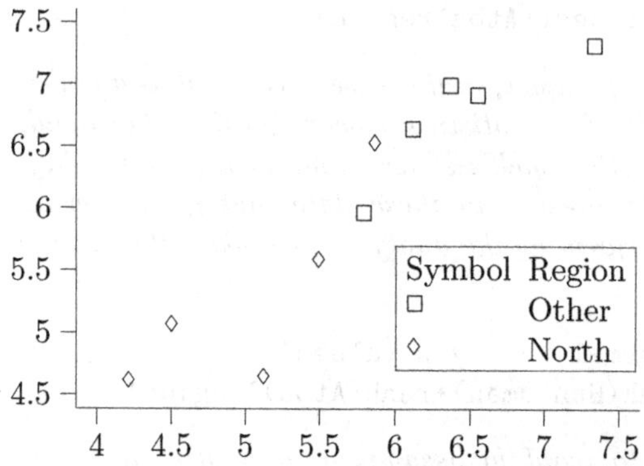

Symbol	Region
□	Other
◇	North

7.5.1.2 Permutation Distribution Approximations

Let T_j be the Mann-Whitney-Wilcoxon statistic using manifest variable j, for $j \in \{1, \ldots, J\}$. Let $\boldsymbol{T} = (T_1, \ldots, T_J)$. Let $\boldsymbol{\Sigma} = \text{Var}[\boldsymbol{T}]$ be the variance matrix of this statistic, under the permutation distribution. Let $\sigma_{jj'}$ be the element in row j and column j'. The diagonal elements σ_{jj} are independent of data values (and equal $M_1 M_2 (M_2 + M_1 + 1)/12$, but that's not important here). The remaining entries of $\boldsymbol{\Sigma}$ depend on the data. For $i = 1, \ldots, M_1 + M_2$, let F_{ij} be the number of observations in group 2 that beat observation i on variable j if i is in group 1, and the number of observations in group 1 that i beats on variable j, if i is in group 2. Then $4/(M_2 + M_1)$ times the covariance matrix for F estimates the variance matrix of $\boldsymbol{T}$. Kawaguchi et al. (2011) provide details of these calculations. Superior performance can be obtained using known diagonal values, and estimated correlations for the remaining entries of the variance matrix (Chen and Kolassa, 2018).

Example 7.5.2 *Consider again the wheat yield data of Example 7.5.1. Asymptotic nonparametric testing is performed using*

```
library(ICSNP)#For rank.ctest and HotelingsT2.
rank.ctest(cbind(Huntsman,Atou)~region)
```

Exercises 141

to obtain a p-value of 0.039, or

```
rank.ctest(cbind(Huntsman,Atou)~region,scores="sign")
detach(wheat)
```

to obtain a multivariate version of Mood's median test.
  Alternate syntax for `rank.ctest` consists of calling it with two arguments corresponding to the two data matrices.

7.6 Exercises

1. The data set

   http://ftp.uni-bayreuth.de/math/statlib/datasets/federalistpapers.txt

   gives data from an analysis of a series of documents. The first column gives document number, the second gives the name of a text file, the third gives a group to which the text is assigned, the fourth represents a measure of the use of first person in the text, the fifth presents a measure of inner thinking, the sixth presents a measure of positivity, and the seventh presents a measure of negativity. There are other columns that you can ignore. (The version at Statlib, above, has odd line breaks. A reformatted version can be found at stat.rutgers.edu/home/kolassa/Data/federalistpapers.txt).

   a. Test the null hypothesis that the multivariate distribution of first person, inner thinking, positivity, and negativity, are the same between groups 1 and 2, using a permutation test. Test at $\alpha = .05$.

   b. Construct new variables, the excess of positivity over negativity, and the excess of thinking ahead over thinking behind, by subtracting variable six minus variable seven, and variable eight minus variable nine. Test the null hypothesis that the multivariate distribution of these two new variables has median zero, versus the general alternative, using the multivariate version of the sign test. Test at $\alpha = .05$.

2. The data at

   http://lib.stat.cmu.edu/datasets/cloud

   contain data from a cloud seeding experiment. The first fifteen lines contain comment and label information; ignore these. The second field contains the character S for a seeded trial, and U for unseeded.

a. The fourth and fifth represent rainfalls in two target areas. Test the null hypothesis that the bivariate distribution of observations after seeding is the same as that without seeding. Use the marginal rank sum test.

b. Repeat part (a) using the permutation version of Hotelling's test.

8
Density Estimation

This chapter considers the task of estimating a density from a sample of independent and identical observations. Previous chapters began with a review of parametric techniques. Parametric techniques for density estimation might involve estimating distribution parameters, and reporting the parametric density with estimates plugged in; this technique will not be further reviewed in this volume.

8.1 Histograms

The most elementary approach to this problem is the histogram, which represents the density as a bar chart.

The bar chart represents frequencies of the values of a set of categorical variable in various categories as the heights of bars. When the categories are ordered, one places the bars in the same order. To construct a histogram for a continuous random variable, then, coarsen the continuous variable into a categorical variable, whose categories are subsets of the range of the original variable. Construct a bar plot for this categorical variable, again, with bars ordered according to the order of the intervals.

Because the choice of intervals is somewhat arbitrary, the boundary between bars is deemphasized by making the bars butt up against their neighbors. The most elementary version of the histogram has the height of the bar as the number of observations in the interval. A more sophisticated analysis makes the height of the bar represent the proportion of observations in the bar, and a still more sophisticated representation makes the area of the bar equal to the proportion of observation in the bar; this allows bars of unequal width while keeping the area under a portion of the curve to approximate the proportion of observation in that region. Unequally-sized bars are unusual.

Construction of a histogram, then, requires selection of bar width and bar starting point. One generally chooses the end points of intervals generating the bars to be round numbers. A deeper question is the number of such intervals. An early argument (Sturges, 1926) involved determining the largest number so that if the data were in proportion to a binomial distribution, every interval in the range would be non-empty. This gives a number of bars proportional to the log of the sample size.

The histogram is determined once the bin width Δ_n, and any bin separation point, is selected. Once Δ_n is selected, many bin separation points determine the same histogram; without loss of generality, choose the smallest non-negative value and denote it by t_n.

Scott (1979) calculates optimal sample size by minimizing mean square error of the histogram as an estimator of the density. Construct the histogram with bars of uniform width, and such that the bar area equals the proportion of observations out of the entire sample falling in the particular bar. Let $\hat{f}(x)$ represents the height of the histogram bar containing the potential data value x. The integrated mean squared error of the approximation is then

$$\int_{-\infty}^{\infty} E\left[|f(x) - \hat{f}(x)|^2\right] dx. \qquad (8.1)$$

The quality of the histogram depends primarily on the width of the bin Δ_n; the choice of t_n is less important. Scott (1979) takes $t_n = 0$, and, using Taylor approximation techniques within the interval, shows that the integrated mean squared error of the approximation is

$$1/(n\Delta_n) + \frac{1}{12}\Delta_n^2 \int_{-\infty}^{\infty} f'(x)^2 \, dx + O(1/n + \Delta_n^3). \qquad (8.2)$$

The first term in (8.2) represents the variance of the estimator, and the second term represents the square of the bias. Minimizing the sum of the first two terms gives the optimal bin size $\Delta_n^* = [6/\int_{-\infty}^{\infty} f'(x)^2 dx]^{1/2} n^{-1/3}$. One might approximate $\int_{-\infty}^{\infty} f'(x)^2 dx$ by using the value of the Gaussian density with variance matching that of the data, to obtain $\Delta_n^* = 3.49 s n^{-1/2}$, for s the standard deviation of the sample.

8.2 Kernel Density Estimates

Histograms are easy to construct, but their choppy shape generally does not reflect our expectations about true density. A more sophisticated approach is kernel density estimation (Rosenblatt, 1956). Estimate the density as the average of densities centered at data points:

$$\hat{f}(x) = (\Delta_n n)^{-1} \sum_{i=1}^{n} w((X_i - x)/\Delta_n). \qquad (8.3)$$

Here w is a density, also called a kernel; that is, a non-negative function integrating to 1.

Kernel Density Estimates

This estimator $\hat{f}$ depends on the kernel w, and the smoothing parameter Δ_n.

Some plausible choices for w are the (standard) Gaussian density, the Epanechnikov kernel $w(x) = a - 16a^3 x^2/9$ for $|x| \leq \frac{3}{4a}$; this kernel is scaled to have unit variance with $a = \frac{3}{4\sqrt{5}}$, and minimizes the integrated mean square error (8.1) among symmetric kernels (Epanechnikov, 1969).

Another plausible kernel is the triangle kernel $w(x) = 1/\sqrt{6} - |x/6|$, for $|x| \leq \sqrt{6}$, and 0 otherwise. One might also consider the box kernel; consider the simplest version, $w(x) = 1$ if $|x| \leq 1/2$, rather than the one standardized to unit variance, $w(x) = \sqrt{3}/2$ if $|x| \leq \sqrt{3}$. This kernel will be considered for pedagogical reasons below, although in practical terms its use abandons the advantages of the smoothness of the result.

Example 8.2.1 *Refer again to the nail arsenic data from Example 2.3.2. Figure 8.1 displays kernel density estimates with a default bandwidth and for a variety of kernels. This figure was constructed using*

```
cat('\n Density estimation   \n')
attach(arsenic)
#Save the density object at the same time it is plotted.
plot(a<-density(nails),lty=1,
   main="Density for Arsenic in Nails for Various Kernels")
lines(density(nails,kernel="epanechnikov"),lty=2)
lines(density(nails,kernel="triangular"),lty=3)
lines(density(nails,kernel="rectangular"),lty=4)
legend(1,2, lty=rep(1,4), legend=c("Normal","Quadratic",
    "Triangular","Rectangular"))
```

Note that the rectangular kernel is excessively choppy, and fits the poorest.

Generally, any other density, symmetric and with a finite variance, may be used. The parameter Δ_n is called the bandwidth. This bandwidth should depend on spread of data, and n; the spread of the data might be described using the standard deviation or interquartile range, or, less reliably, the sample range.

The choice of bandwidth balances effects of variance and bias of the density estimator, just as the choice of bin width did for the histogram. If the bandwidth is too high, the density estimate will be too smooth, and hide features of data. If the bandwidth is too low, the density estimate will provide too much clutter to make understanding the distribution possible.

We might consider minimizing the mean squared error for x fixed, rather than after integration. That is, choose Δ_n to minimize the mean square error of the estimate,

$$\mathrm{MSE}[\hat{f}(x)] = \mathrm{Var}\left[\hat{f}(x)\right] + \left(\mathrm{E}\left[\hat{f}(x)\right] - f(x)\right)^2.$$

FIGURE 8.1: Density for Arsenic in Nails for Various Kernels

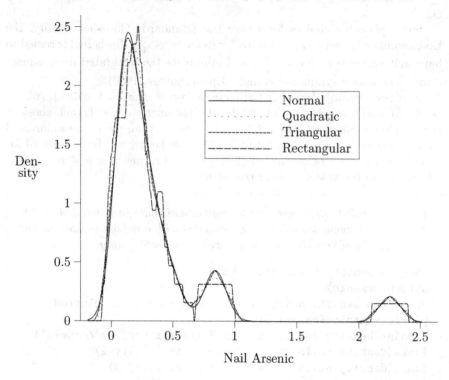

Using the box kernel, the estimate $\hat{f}(x)$ has a rescaled binomial distribution, and $\mathrm{E}\left[\hat{f}(x)\right] = p/\Delta_n$ for $p = F(x+\Delta_n/2) - F(x-\Delta_n/2)$. Expanding $F(x)$ as a Taylor series, noting that $F'(x) = f(x)$, and canceling terms when possible,

$$p/\Delta_n = f(x) + \Delta_n^2 f''(x^*)/24 \text{ for some } x^* \in [x - \Delta_n/2, x + \Delta_n/2]. \quad (8.4)$$

The bias of the estimator is approximately $\Delta_n^2 f''(x^*)/24$. Hence, generally, if Δ_n does not converge to zero as $n \to \infty$, then bias does not converge to zero, and the estimate is inconsistent; hence consider only strategies for which $\lim_n \Delta_n = 0$. Furthermore, since $\lim_n \Delta_n = 0$, then bias is approximated by

$$\Delta_n^2 C_1(x) \text{ for } C_1(x) = f''(x^*)/24. \quad (8.5)$$

Furthermore, $\mathrm{Var}\left[\hat{f}(x)\right] = p(1-p)/(n\Delta_n^2)$, and applying (8.4),

$$\mathrm{Var}\left[\hat{f}(x)\right] = \frac{\left(1 - \Delta_n f(x) - \frac{1}{24}\Delta_n^3 f''(x^*)\right)\left(f(x) + \frac{1}{24}\Delta_n^2 f''(x^*)\right)}{\Delta_n n},$$

and using the convergence of Δ_n to zero, one obtains

$$\mathrm{Var}\left[\hat{f}(x)\right] \approx C_2(x)/(\Delta_n n) \text{ for } C_2(x) = f(x). \quad (8.6)$$

Kernel Density Estimates

Minimizing the mean square requires minimizing

$$C_2(x)/(\Delta_n n) + C_1(x)^2 \Delta_n^4. \tag{8.7}$$

Differentiating and setting (8.7) to zero implies that $C_2(x)\Delta_n^{-2}/n = 4C_1(x)^2\Delta_n^3$, or

$$\begin{aligned}\Delta_n &= 2^{-2/5}C_2(x)^{1/5}C_1(x)^{-2/5}n^{-1/5} \\ &= f(x)^{1/5}2^{-2/5}(f''(x))^{-2/5}n^{-1/5}.\end{aligned} \tag{8.8}$$

This gives a bandwidth dependent on x; a bandwidth independent of x may be constructed by minimizing the integral of the mean squared error (8.7), $C_2/(\Delta_n n) + C_1^2 \Delta_n^4$, for $C_2 = \int_{-\infty}^{\infty} C_2(x)\, dx$ and $C_1 = \sqrt{\int_{-\infty}^{\infty} C_1^2(x)\, dx}$.

Example 8.2.2 *Refer again to the arsenic nail data of Examples 2.3.2 and 8.2.1. The kernel density estimate, using the suggested bandwidth, and bandwidths substantially larger and smaller than the optimal, are given in Figure 8.2. The excessively small bandwidth follows separate data points closely, but obscures the way they cluster together. The excessively large bandwidth wipes out all detail from the data set. Note the large probability assigned negative arsenic concentrations by the estimate with the excessively large bandwidth. This plot was drawn in R using*

```
plot(density(nails,bw=a$bw/10),xlab="Arsenic",type="l",
    main="Density estimate with inopportune bandwidths")
lines(density(nails,bw=a$bw*5),lty=2)
legend(1,2,legend=paste("Band width default",c("/10","*5")),
    lty=c(1,2))
detach(arsenic)
```

and the object a *is the kernel density estimate with the default bandwidth, constructed in Example 8.2.1.*

Silverman (1986, §3.3) discusses a more general kernel $w(t)$. Equations (8.5) and (8.6) hold, with

$$C_1(x) = f(x)\int_{-\infty}^{\infty} w(t)^2\, dt,\ C_2(x) = 1/2 f''(x)\int_{-\infty}^{\infty} t^2 w(t)\, dt,$$

and (8.8) continues to hold. Sheather and Jones (1991) further discuss the constants involved.

Epanechnikov (1969) demonstrates that (8.3) extends to multivariate distributions; estimate the multivariate density $f(\boldsymbol{x})$ from independent and identically distributed observations $\boldsymbol{X}_1, \ldots, \boldsymbol{X}_n$, where $\boldsymbol{X}_i = (X_{i1}, \ldots, X_{id})$, using

$$\hat{f}(\boldsymbol{x}) = n^{-1}(\prod_{j=1}^{d} \Delta_{nj})^{-1}\sum_{i=1}^{n} w((X_{i1}-x_1)/\Delta_{n1},\ldots,(X_{id}-x_d)/\Delta_{nd}).$$

FIGURE 8.2: Density Estimate with Inopportune Bandwidths

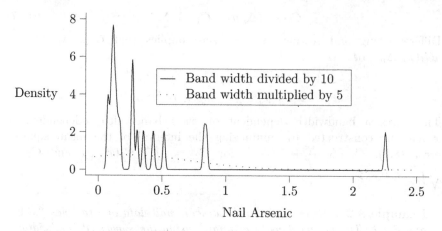

Here w is a density over $\Re^d$, and Δ_{nj} are dimension-dependent bandwidth parameters. Epanechnikov (1969) uses multivariate densities that are products of separate densities for each dimension, but surveys work using more general kernels in two dimensions. The number of observations needed for precise estimation grows dramatically as the dimension d increases.

8.3 Exercises

1. The data at

    http://lib.stat.cmu.edu/datasets/CPS_85_Wages

    reflects wages from 1985. The first 27 lines of this file represent an explanation of variables; delete these lines first, or skip them when you read the file. The first six fields are numeric. The sixth is hourly wage; you can skip everything else. Fit a kernel density estimate to the distribution of wage, and plot the result. Comment on what you see.

2. The data at

    http://lib.stat.cmu.edu/datasets/cloud

    contain data from a cloud seeding experiment. The first fifteen lines contain comment and label information; ignore these. The third field indicates season, and the sixth field represents a rainfall measure expected not to be impacted by the experimental intervention. On the same set of axes, plot kernel density estimates of summer and winter rain fall, and comment on the difference.

9
Regression Function Estimates

Consider modeling a collection of responses Y_j as a function of explanatory variables $\boldsymbol{X}_j = (X_{j1}, \ldots, X_{jK})$. That is, express

$$Y_j = g(\boldsymbol{X}_j) + \epsilon_j, \tag{9.1}$$

with errors ϵ_j independent and typically standardized to have a common scale measure such as standard deviation or interquartile range, and with various constraints on g. In this chapter, assume further that these errors are identically distributed. The least restrictive constraint on g allows for an arbitrary conditional expectation of Y_j; since in many cases each unique value of X_j appears only a small number of times in the data set, fitting such a model reliably is difficult.

As with the previous chapters (except for Chapter 8), this chapter begins with a review of standard Gaussian-theory results. Unlike previous chapters, it also includes, in the form of quantile and resistant regression techniques, some methods that are partially parametric in nature.

9.1 Standard Regression Inference

The most restrictive constraint on g is to require an affine function of $\boldsymbol{X}_j$; that is,

$$g(\boldsymbol{X}_j) = \boldsymbol{\beta}^\top \boldsymbol{X}_j; \tag{9.2}$$

here $\boldsymbol{\beta}$ and $\boldsymbol{X}_j$ are column vectors. Generally, the first component of each of the $\boldsymbol{X}_j$ is 1, making the first component of $\boldsymbol{\beta}$ an intercept parameter.

The notation above, with $\boldsymbol{X}_i$ capitalized, implies that the explanatory variables are random. Analyses considered in this chapter are generally performed conditionally on these response variables, and so their random nature might be ignored.

The standard approach chooses the regression parameters to minimize the sum of squares of residuals; that is,

$$\hat{\boldsymbol{\beta}} = \underset{\boldsymbol{\beta}}{\operatorname{argmin}} \sum_{i=1}^{n} (Y_i - \boldsymbol{\beta}^\top \boldsymbol{X}_i)^2 = (\boldsymbol{X}^\top \boldsymbol{X})^{-1} \boldsymbol{X}^\top \boldsymbol{Y}, \tag{9.3}$$

for

$$\boldsymbol{X} \text{ the } n \times K \text{ matrix with rows given by } \boldsymbol{X}_i^\top, \quad (9.4)$$

and $\boldsymbol{Y}$ is the column vector with entries Y_i. The estimator (9.3) is defined only if $\boldsymbol{X}$ is of full rank, or, in other words, if the inverse in (9.3) exists. Otherwise, the model is not identifiable, in that two different parameter vectors $\boldsymbol{\beta}$ and $\boldsymbol{\beta}^*$ give the same fitted values, or $\boldsymbol{X\beta} = \boldsymbol{X\beta}^*$.

When the errors ϵ_j have a Gaussian distribution, then the vector $\hat{\boldsymbol{\beta}}$ has a multivariate Gaussian distribution, exactly, and

$$(\hat{\beta}_i - \beta_i)/\sqrt{s^2 v_i} \sim \mathfrak{T}_{n-K}, \quad (9.5)$$

for $s^2 = \sum_{i=1}^n (Y_i - \boldsymbol{\beta}^\top \boldsymbol{X}_i)^2/(n-K)$, and v_i the entry in row and column i of $(\boldsymbol{X}^\top \boldsymbol{X})^{-1}$. When the errors ϵ_j are not Gaussian, under certain circumstances, a central limit theorem justifies approximating the distribution of $\hat{\boldsymbol{\beta}}$ as multivariate Gaussian, and (9.5) holds, approximately.

9.2 Kernel and Local Regression Smoothing

An intermediate level of constraint has $g(\boldsymbol{x})$ continuous and differentiable, with curves that turn quickly discouraged. One method of fitting under such a constraint is kernel smoothing, and specifically as Nadaraya-Watson smoothing (Nadaraya, 1964; Watson, 1964). One obtains an expression that is explicit rather than implicit; estimate $g(x)$ as

$$\hat{g}(x) = \sum_{j=1}^n Y_j w((\boldsymbol{x} - \boldsymbol{X}_j)/\Delta_n) / \sum_{j=1}^n w((\boldsymbol{x} - \boldsymbol{X}_j)/\Delta_n). \quad (9.6)$$

The weight function can be the same as was used for kernel density estimation. This weight function is often a Gaussian density, or a uniform density centered at 0. Fan (1992) discusses a local regression smoother

$$\hat{g}(x) = \sum_{\ell=0}^L \hat{\beta}_\ell x^\ell, \quad (9.7)$$

for $L=1$ and

$$\hat{\boldsymbol{\beta}} = \operatorname{argmin}\left(\sum_{j=1}^n \left(Y_j - \sum_{\ell=0}^L \hat{\beta}_\ell X_j^\ell \right)^2 w((x - X_j)/\Delta_n) \right), \quad (9.8)$$

and argues that this estimator has smaller bias than (9.6). Köhler et al. (2014) present considerations for bandwidth selection.

Example 9.2.1
Example 2.3.2 presents nail arsenic levels from a sample; arsenic levels in drinking water were also recorded, and one can investigate the dependence of nail arsenic on arsenic in water. The data may be plotted in R using

```
attach(arsenic)
plot(water,nails,main="Nail and Water Arsenic Levels",
   xlab="Water",ylab="Nails",
   sub="Smoother fits.  Bandwidth chosen by inspection.")
```

Kernel smoothing may be performed in R using the function ksmooth. *Bandwidth was chosen by hand.*

```
lines(ksmooth(water,nails,"normal",bandwidth=0.10),lty=1)
lines(ksmooth(water,nails,"box",bandwidth=0.10),lty=2)
legend(0.0,2.0,lty=1:3, legend=c("Smoother, Normal Kernel",
   "Smoother, Box Kernel", "Local Polynomial"))
```

The function locpoly *from library* KernSmooth *applies a local regression smoother:*

```
library(KernSmooth)
lines(locpoly(water,nails,bandwidth=.05,degree=1),lty=3)
detach(arsenic)
```

Results are given in Figure 9.1. The box kernel performs poorly; a sample this small necessarily undermines smoothness of the box kernel fit. Perhaps the normal smoother is under-smoothed, but not by much. Library KernSmooth *contains a tool* dpill *for automatically selecting bandwidth. The documentation for this function indicates that it sometimes fails, and, in fact,* dpill *failed in this case. This local regression smoother ignored the point with the largest values for each variable, giving the curve a concave rather than convex shape.*

These data might have been jointly modeled on the square-root scale, to avoid issues relating to the distance of the point with the largest values for each variable from the rest of the data. An exercise suggests exploring this; in this case, the automatic bandwidth selector dpill *returns a value.*

Figure 9.2 demonstrates the results of selecting a bandwidth too small. In this case, the Gaussian kernel gives results approximately constant in the neighborhood of each data point, and the box kernel result is not defined for portions of the domain, because both numerator and denominator in (9.6) are zero.

FIGURE 9.1: Nail Arsenic and Water Arsenic

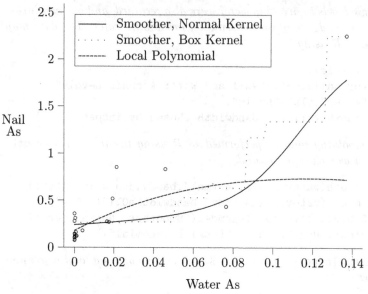

Smoother fits. Bandwidth chosen by inspection.

FIGURE 9.2: Nail Arsenic and Water Arsenic

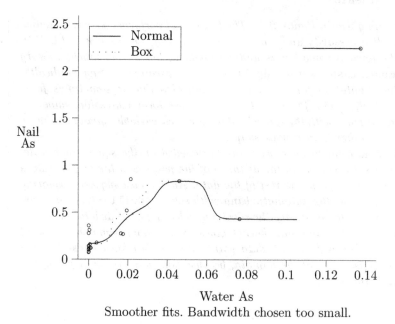

Smoother fits. Bandwidth chosen too small.

Kernel and Local Regression Smoothing

An alternative procedure uses as $\hat{g}(x)$ the fitted value at x for low-degree (viz., linear or quadratic) regression of points with X_j near x (Cleveland, 1979; Savitzky and Golay, 1964). Cleveland and Devlin (1988) refer to this procedure as locally weighted regression (loess). One specifies the number of points k, and up-weights points near x and down-weights them away from x. The weighting function is scaled to make the point in the neighborhood farthest from x have a weight going down to zero.

Consider the case with an intercept and one regressor, so $L = 2$. The estimate is now (9.7), for

$$\hat{\beta} = \operatorname{argmin}\left(\sum_{j \in N(x)} \left(Y_j - \sum_{\ell=0}^{L} \beta_\ell X_j^\ell\right)^2 w((x - X_j)/\Delta_n(x))\right), \quad (9.9)$$

for $N(x)$ the indices of the k closest points to x, and

$$\Delta_n(x) = \max\{|X_j - x| | j \in N(x)\}. \quad (9.10)$$

The local linear regression fit (9.8) with (9.6) differs from (9.9) with (9.6) in the restriction of the consideration of points defining local regression parameter estimates to points near the point of interest, rather than using all points with positive values for a weighting function. Note further that for the loess procedure, bandwidth is determined implicitly, and depends on the point at which the smoother is applied.

A common weight function is $w(x) = (1 - |x|^3)^3$.

Example 9.2.2 *The solution using the default proportion of the data 0.75 appears to under-smooth the data, and raising this parameter above 1 reduces curvature, but not by much. Return again to the arsenic data previously examined in Example 9.2.1. Again, plot the data in R using*

```
attach(arsenic)
plot(water,nails,main="Nail and Water Arsenic Levels",
   xlab="Water",ylab="Nails", sub="Loess fits")
```

Loess smoothing may be performed in R using the function `loess`. *Bandwidth was chosen by hand. Unlike in the case of* `ksmooth`, `loess` *does not provide output that can be plotted directly. A set of points at which to calculate the smoother must be specified; this is stored in the variable* `x` *below. Instead of specifying a bandwidth for the loess procedure, one specifies the number of observations contributing to each neighborhood from which the fit is calculated. In R, this is specified as the proportion of the total sample, via the input parameter* `span`. *Hence the second call below uses the entire data set.*

```
x<-min(water)+(0:50)*diff(range(water))/50
lines(x,y=predict(loess(nails~water),x))
```

```
lines(x,y=predict(loess(nails~water,span=1),x),lty=2)
```

Values of span *above 1 specify for the bandwidth to be expanded beyond (9.10), thus increasing smoothing:*

```
lines(x,y=predict(loess(nails~water,span=10),x),lty=3)
legend(0.0,2.0,legend=c("Default span .75",
   "Span 1","Span 10"),lty=1:3)
detach(arsenic)
```

Results are given in Figure 9.3. Note that the solution using the default proportion of the data 0.75 appears to under-smooth the data, and that raising this parameter above 1 misses detail of the shape of the relationship.

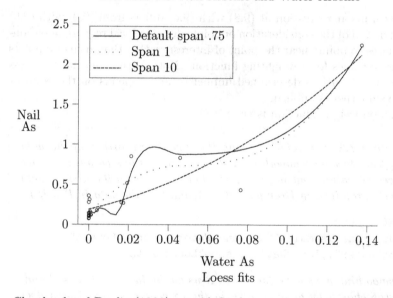

FIGURE 9.3: Nail Arsenic and Water Arsenic

Loess fits

Cleveland and Devlin (1988) extend this loess technique to higher dimensions.

9.3 Isotonic Regression

Many contexts justify a non-decreasing or non-increasing nonparametric relationship between variables. In this section consider non-decreasing relationships; reversing this constraint is straight-forward.

Splines

In the case in which $K=1$, one might choose $\hat{Y}_j$ to minimize $\sum_{j=1}^{n}(Y_j - \hat{Y}_j)^2$ subject to $\hat{Y}_j \geq \hat{Y}_i$ whenever $X_j \geq X_i$. This model fitting is an example of quadratic programming. When $K > 1$, then the space of possible covariates does not have a natural ordering. One may construct a partial ordering; that is, for some distinct covariate vectors x and y, neither x nor y is ordered higher. For example, one may define the partial ordering $(x_1, \ldots, x_K) \preceq (y_1, \ldots, y_K)$ if $x_i \preceq y_i$ for all i; that is, consider one vector as greater than or equal to the other if and only if each component of the first vector does not exceed the same component of the second vector. Such an ordering lacks the property that any two elements of the set may be compared, and so is called a partial ordering. For example, neither $(2,1) \preceq (1,2)$ nor $(1,2) \preceq (2,1)$. In the case of a single regressor this complication does not arise, since for any two distinct real numbers, one can be determined to be the smaller and the other is the larger.

Brunk (1955) introduces the notion of model fitting that respects the partial ordering $g(x) \leq g(y)$ if $x \preceq y$. Such techniques are called isotonic regression. Dykstra (1981) reviews an algorithm for fitting such a model, called the pooled adjacent violators algorithm, and produces theoretical justification for this algorithm. Best and Chakravarti (1990) reviews more general algorithmic considerations.

Example 9.3.1 *Example 2.3.2 presents arsenic levels in water and nails. Isotonic regression may be performed in R using the function* isoreg, *and R provides a plotting method for isotonic regression results. Hence the regression output may be plotted directly, using*

```
attach(arsenic)
plot(isoreg(water,nails), xlab="Water As",ylab="Nail As",
    main="Nail Arsenic and Water Arsenic")
detach(arsenic)
```

Results are given in Figure 9.4.

9.4 Splines

A spline is a smooth curve approximating the relationship between an explanatory variable x and a response variable y, based on observed pairs of points $(X_1, Y_1), \ldots, (X_n, Y_n)$, constructed according to the following method, in order to describe the dependence of y on x between two points x_0 and x_N. One first picks $N-1$ intermediate points $x_1 < x_2 < \cdots < x_{N-2} < x_{N-1}$. The intermediate points are called knots. One then determines a polynomial of degree M between x_{j-1} and x_j, constrained so that the derivatives of order up to $M-1$ match up at knots. Denote the fitted mean by $\hat{g}(x)$.

FIGURE 9.4: Nail Arsenic and Water Arsenic

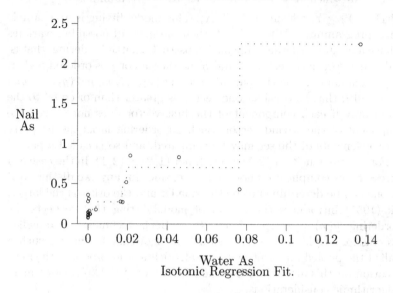

Isotonic Regression Fit.

Taken to an extreme, if all X_j are unique, then one can fit all n points with a polynomial of degree $n-1$; this, however, will yield a fit with unrealistically extreme fluctuations. Instead, choose the polynomials to minimize

$$\sum_{j=1}^{n}(Y_j - \hat{g}(X_j))^2 + \lambda \int_{X_{(1)}}^{X_{(n)}} |\hat{g}''(x)|\ dx;$$

here λ is a parameter that penalizes estimates with large curvature.

Example 9.4.1 *Again revisit the arsenic data of Example 9.2.1. The spline is fit using R as*

```
attach(arsenic)
plot(water,nails, main="Nail Arsenic and Water Arsenic")
hgrid<-min(water)+(0:50)*diff(range(water))/50
lines(predict(spl<-smooth.spline(water,nails),hgrid))
lines(predict(smooth.spline(water,nails,spar=.1),hgrid),
    lty=3)
lines(predict(smooth.spline(water,nails,spar=.5),hgrid),
    lty=2)
legend(1250,1110,lty=1:3,col=1:3,
    legend=paste("Smoothing",c(round(spl$spar,3),.5,.1),
    c("Default","","")))
detach(arsenic)
```

> Here `hgrid` is a set of points at which to calculate the spline fit. As with most of the smoothing methods, R contains a lines method that will plot predicted values directly. See Figure 9.5. Smoothing parameters smaller than optimal yield over-fitted curves.

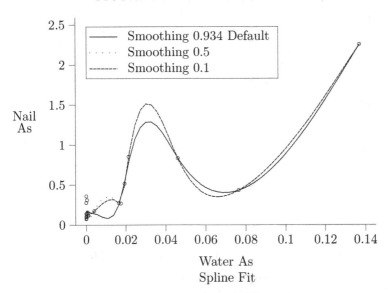

FIGURE 9.5: Nail Arsenic and Water Arsenic

Schoenberg (1946) introduced this technique.

9.5 Quantile Regression

Least squares regression, as in (9.3), provides parameter estimates minimizing the sum of squares of errors in fitting. Such an approach is often criticized as allowing excessive influence to outliers, and, hence, is non-robust. As an alternative, consider minimizing the sum of residuals, to obtain

$$\hat{\beta} = \operatorname*{argmin}_{\beta} \left[\sum_{j=1}^{n} |Y_j - \beta_1 - \beta_2 X_j| \right]. \tag{9.11}$$

When comparing (9.11) with the standard least-squares criterion (9.3), note the absence of $\cdot^2$ in (9.11).

Equivalently, one might minimize

$$\sum_j e_j^+ + e_j^-, \tag{9.12}$$

for
$$e_j^+ \geq 0,\ e_j^- \geq 0,\ e_j^+ - e_j^- = Y_j - \beta_1 - \beta_2 X_j \forall j. \qquad (9.13)$$

The objective function is not differentiable; the optimum is where (9.12) is either at a point, or flat, making the optimizer not unique.

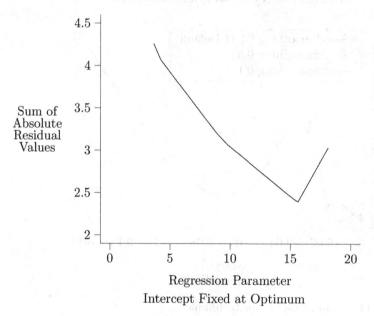

FIGURE 9.6: Hypothetical L^1 Fit

This solution is called <u>L^1 regression</u>, after the power on $|Y_j - \beta_1 - \beta_2 X_j|$. This solution is also called <u>quantile regression</u>. To understand the motivation behind this term, note that if $\beta_2 = 0$, then the best β_1 is the median. More generally, if the linear model fits, and if the errors are identically distributed, then the quantile regression line runs through the median of the distribution of the responses, conditional on the explanatory variable. Note the parallelism between the minimizations of (2.4) and (9.11). Just as median is not always uniquely defined, these estimates are not necessarily uniquely defined.

Stigler (1984) attributes the suggestion to fit the linear model using (9.12) to Roger Boscovich, and a partial solution to the minimization problem to Thomas Simpson, both circa 1760. Stigler (1973) attributes the solution of the problem when $\beta_1 = 0$ to Laplace (1818). Koenker (2000) traces further developments through the work of Edgeworth.

Table 9.1 compares coverage and length for the quantile and least squares regression estimators, for a variety of distributions and sample sizes. Surprisingly, both techniques maintain approximate 90% coverage for all putative distributions and all sample sizes. Unsurprisingly, L^2 intervals are shorter for Gaussian variables, uniform variables, and for Laplace and exponential vari-

TABLE 9.1: Achieved coverage and average interval length for 90% confidence intervals for L^1 and L^2 regression, with various error distributions

	Interval Coverage, 10 observations				
	Gaussian	Cauchy	Laplace	Uniform	Exponential
L^1	0.8854	0.8783	0.8811	0.8830	0.8826
L^2	0.9051	0.9179	0.9042	0.8975	0.9047
	Average Length, 10 observations				
	Gaussian	Cauchy	Laplace	Uniform	Exponential
L^1	0.4766	3.5869	0.6092	0.1441	0.4035
L^2	0.3981	10.8032	0.5466	0.1162	0.3773
	Interval Coverage, 50 observations				
	Gaussian	Cauchy	Laplace	Uniform	Exponential
L^1	0.8876	0.8832	0.8847	0.8876	0.8900
L^2	0.9006	0.9121	0.9047	0.9017	0.9021
	Average Length, 50 observations				
	Gaussian	Cauchy	Laplace	Uniform	Exponential
L^1	0.0391	0.0568	0.0386	0.0145	0.0303
L^2	0.0363	2.0033	0.0509	0.0105	0.0357

ables with the smaller sample size, and longer for Cauchy variables and for Laplace and exponential variables with the larger sample size.

One can adapt this technique to fit quantiles of the errors other than the median. One can replace the objective (9.12) by

$$\gamma \sum_j e_j^+ + (1-\gamma) \sum_j e_j^-, \qquad (9.14)$$

still subject to constraints (9.13). This change causes the regression line to run through the $1-\gamma$ quantile of $Y|X$.

9.5.1 Fitting the Quantile Regression Model

This minimization of (9.12) or (9.14), subject to (9.13), is an example of linear programming. The solution to this minimization is more computationally intensive than the minimization of (9.3), and, unlike the least squares case, does not have a closed-form expression. This added difficulty arises because the objectives (9.12) and (9.14) are non-differentiable. At the minimizer $\hat{\beta}$, the objective function does not decrease as β moves away from $\hat{\beta}$. Since $\sum_{j=1}^n |Y_j - \beta_1 - \beta_2 X_j|$ is a piecewise-linear function of β_1 and β_2, the function $S(\beta)$ giving a counterpart of the derivative is piecewise constant, with jumps. Then $\hat{\beta}$ should satisfy $S(\hat{\beta}) \approx 0$. The estimate either sets the function to 0, or is a point where it jumps across 0. This score function $S(\beta)$ may be expressed as a rank statistic, and the null hypothesis $H_0 : \beta = \beta^\circ$ may be tested by comparing $S(\beta°)$ to $\mathbf{0}$, either exactly or asymptotically. This test can be inverted to give confidence sets for β.

A solution to the more general problem seems to have been an early application of general linear programming methods.

Example 9.5.1 *Consider again the arsenic data of Example 9.2.1. The quantile regression fitter is found in R in library* **quantreg**. *The following commands fit this model:*

```
library(quantreg)#Need for rq
rq(arsenic$nails~arsenic$water)
```

to give the results

```
Call: rq(formula = nails ~ water)

tau: [1] 0.5

Coefficients:
            coefficients  lower bd  upper bd
(Intercept)   0.11420      0.10254   0.14093
water        15.60440      3.63161  18.07802
```

Hence the best estimate of the slope of the linear relationship of arsenic in nails to arsenic in water is 15.604, with 90% confidence interval (3.632,18.078). This confidence level is the default for **rq**. *Figure 9.6 is drawn using*

```
attach(arsenic)
rqo<-summary(rq(nails~water))$coef
regp<-rqo[2,2]+diff(rqo[2,2:3])*(0:100)/100
m<-apply(abs(outer(nails,rep(1,length(regp)))
   - outer(water,regp)-rqo[1,1]),2,sum)
plot(regp,m,type="l",xlab="Regression Parameter",
    ylab="Sum of absolute values of residuals",
    main="L1 fit for Arsenic Large Scale",
    sub="Intercept fixed at optimum")
detach(arsenic)
```

and demonstrates the estimation of the slope; in this figure the intercept is held at its optimal value.

Compare this to the Theil-Sen estimator:

```
attach(arsenic)
library(deming); theilsen(nails~water,conf=.9)$coef
```

yielding

Quantile Regression

> (Intercept)   water
>   0.1167158   14.2156504
>
> *A confidence interval for the slope parameter is calculated using*
>
> `library(MultNonParam);theil(water,nails)`
> `detach(arsenic)`
>
> *giving the confidence interval (8.12,16.41); note the higher precision of this approach.*

If the linear model fits, and if the variance of the errors does not depend on the explanatory variable, then the lines representing various quantiles of the distribution of the response conditional on the explanatory variable will be parallel, but this parallelism need not hold for the estimates in any particular data set, as can be seen in the next example.

> **Example 9.5.2** *Consider the blood pressure data set of Example (6.4.2). Quantile regression can be used to model the median systolic blood pressure after treatment in terms of systolic blood pressure before treatment.*
>
> `#Expect warnings about nonunique estimators. This is OK.`
> `rqout<-rq(bp$spa~bp$spb); summary(rqout)`
>
> *yielding*
>
> ```
>              coefficients  lower bd   upper bd
> (Intercept)   -11.41026    -34.76720  29.00312
> bp$spb          0.94872      0.69282   1.11167
> ```
>
> *Hence blood pressure after treatment increases with blood pressure before treatment; the best estimate indicates that a one-for-one increase is plausible, in that 1 sits inside the confidence interval. One can also fit the .2 quantile:*
>
> `rqoutt<-rq(bp$spa~bp$spb,tau=.2); summary(rqoutt)`
>
> *to obtain*
>
> ```
> Coefficients:
>              coefficients  lower bd   upper bd
> (Intercept)    17.25000    -68.48262  23.34294
> bp$spb          0.75000      0.40129   1.20640
> ```
>
> *One can plot these relationships:*
>
> `plot(bp$spb,bp$spa,main="Systolic Blood Pressure",`
> `    xlab="Before Treatment",ylab="After Treatment")`

```
abline(rqout);abline(rqoutt,lty=2)
legend(150,200,legend=c("Median Regression",
    ".2 Quantile Regresson"),lty=1:2)
```

Results are plotted in Figure 9.7. Note the extreme lack of parallelism.

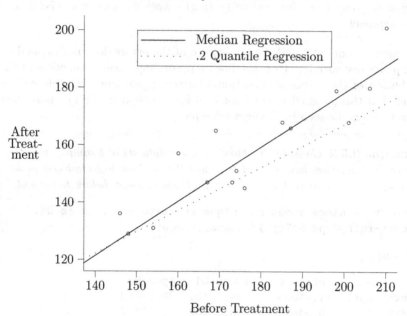

FIGURE 9.7: Systolic Blood Pressure

9.6 Resistant Regression

Again consider the linear model (9.1) and (9.2), with an estimator of the form

$$\hat{\boldsymbol{\beta}} = \underset{\boldsymbol{\beta}}{\operatorname{argmin}} \left[\sum_{i=1}^{n} \varrho((Y_i - \boldsymbol{\beta}^\top \boldsymbol{x}_i)/\sigma) \right]. \tag{9.15}$$

This extends the location estimator (2.6). Least-squares estimates are given by $\varrho(z) = z^2/2$, and the estimator of §9.5 is given by $\varrho(z) = |z|$. These were the same penalty functions used in §2.3.1 to give the sample mean and median respectively. In both of these cases, the regression parameters minimizing the penalty function did not depend on σ.

In parallel with that analysis, one might form an intermediate technique between ordinary and quantile regression by solving (9.15) for an alternative choice of ϱ. One might use (2.7); in such a case, one begins with an

Resistant Regression

initial estimate of σ, and minimization of (9.15) is typically alternated with re-estimation of σ via the method of moments in a recursive manner. This produces estimates that are <u>resistant</u> to the effects of outliers.

Example 9.6.1 *Consider again the blood pressure data of Example 6.4.2. Model the systolic blood pressure after treatment as linear on the blood pressure before treatment, and fit the parameters using the least-squares estimator (9.3), the quantile regression estimator (9.11), and general estimator (9.15) using the resistant penalty (2.7).*

```
library(MASS); rlmout<-rlm(spa~spb,data=bp)
library(quantreg); rqout<-rq(spa~spb,data=bp)
plot(bp$spb,bp$spa,
    main="Blood Pressure After and Before Treatment",
    xlab="Systolic Blood Pressure Before Treatment (mm HG)",
    ylab="Systolic Blood Pressure After Treatment (mm HG)",
    sub="Linear Fits to Original Data")
abline(rlmout,lty=2)
abline(rqout,lty=3)
abline(lm(spa~spb,data=bp),lty=4)
legend(min(bp$spb),max(bp$spa),
 legend=c("Huber","Median","OLS"),lty=2:4)
```

Compare this with results for a contaminated data set. Change the response variable for the first observation to an outlier, and refit.

```
bp$spa[1]<-120
rqout<-rq(bp$spa~bp$spb);rlmout<-rlm(bp$spa~bp$spb)
plot(bp$spb,bp$spa,
    xlab="Systolic Blood Pressure Before Treatment (mm HG)",
    ylab="Systolic Blood Pressure After Treatment (mm HG)",
    sub="Linear Fits to Perturbed Data")
abline(rlmout,lty=2);abline(rqout,lty=3);
abline(lm(spa~spb,data=bp),lty=4)
legend(min(bp$spb),max(bp$spa),
 legend=c("Huber","Median","OLS"),lty=2:4)
```

These fits are shown in Figure 9.8. The median regression line is slightly different in the two plots, although if one does not look closely one might miss this. The least squares fit is radically different as a result of the shift in the data point. The Huber fit is moved noticeably, but not radically.

FIGURE 9.8: Blood Pressure after and before Treatment, Original and Perturbed Data

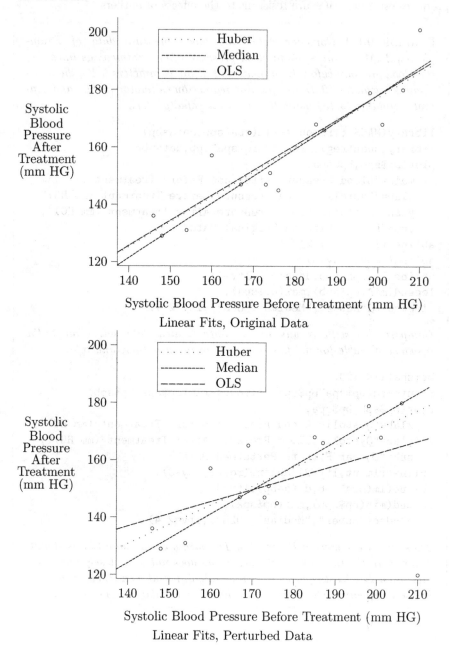

9.7 Exercises

1. The data at

    http://lib.stat.cmu.edu/datasets/CPS_85_Wages

    reflects wages from 1985. The first 42 lines of this file contain a description of the data set, and an explanation of variables; delete these lines first, or skip them when you read the file. The first six fields are numeric. The first is educational level, and the sixth is hourly wage; you can skip everything else.

    a. Fit a relationship of wage on educational level, in such a way as to minimize the sum of absolute value of residuals. Plot this, and compare with the regular least squares regression.

    b. Fit a relationship of wage on educational level, in such a way as to enforce monotonicity. Superimpose a plot of this relationship on a plot of the data.

    c. Fit a kernel smoother to estimate the dependence of wage on educational level, and compare it to the results for the loess smoother.

    d. Fit a smoothing spline to estimate the dependence of wage on educational level, and compare it to the results for the loess smoother and the kernel smoother.

2. Repeat the analysis of Example 9.2.1 for arsenic levels on the square root scale. As an intermediate step, calculate the optimal bandwidth for the local linear smoother. Plot your results. Compare with the results in Figure 9.1.

10

Resampling Techniques

This chapter addresses general procedures for performing statistical inference on a parameter using an estimator with minimal assumptions about its distribution. In contrast to fully nonparametric approaches used earlier, these techniques use information contained in the sample to make inferences about the sampling distribution of the estimator. Two approaches, the bootstrap and the jackknife, are discussed in this chapter.

Suppose that independent and identically distributed vector observations $Z_1, \ldots, Z_n$ are drawn from a distribution with a continuous cumulative distribution function F, and suppose that inference on a parameter θ is required, and an estimator $\hat{\theta}(Z_1, \ldots, Z_n)$ is used.

The parameter θ can be thought of as a function of the distribution function; for example, the population expectation $\theta(F) = \int z \, dF(z)$, and the population median solves $F(\theta) = 1/2$. Furthermore, the estimator $\hat{\theta}$ is a function of the empirical distribution function $\hat{F}$, and often is the same functional that gives the parameter: $\hat{\theta} = \theta(\hat{F})$. For example, an estimator of a population expectation is the sample mean, which is the expectation of the population formed by placing equal weights on each sample point. As a second example, an estimator of a population median is the sample median, which is the median of population formed by placing equal weights on each sample point.

The distribution of the estimator, and particularly how this distribution depends on the quantity to be estimated, is also required for statistical inference. In some cases, the structure of the model for a data set implies the distribution of the estimator; for example, if observations are stochastically independent indicators of whether an event occurs, the resulting distribution is known to be derivable from the distribution of Bernoulli trials. In other cases, recourse is made to a central limit theorem, to produce traditional Gaussian-theory inference. The current chapter concerns using the the observed data set to provide this distributional information.

10.1 The Bootstrap Idea

The bootstrap is a suite of tools for inference on a model parameter, while using the data to give information that might otherwise come from assumptions about the parametric shape of its distribution. Let $G(\hat{\theta}; F)$ represent the desired, but unobservable, distribution of $\hat{\theta}$ computed from independent random vectors $Z_1, \ldots, Z_n$, with each random vector Z_i having a distribution function F. Assume that this distribution depends on F only via the parameter of interest, so that one might write $G(\hat{\theta}; F) = H(\hat{\theta}; \theta)$ for some function H.

Consider constructing confidence intervals using the argument of §1.2.2.1. Using (1.18) with $T = \hat{\theta}$, a $1 - \alpha$ confidence interval for θ satisfies $H(\hat{\theta}; \theta_L) = 1 - \alpha/2$ and $H(\hat{\theta}; \theta_U) = \alpha/2$. The function H, as a function of $\hat{\theta}$, is unknown, and will be estimated from the data. Since observed data all arise from a distribution governed by a single value of θ, the dependence of $H(\hat{\theta}; \theta)$ on θ cannot be estimated. An assumption is necessary in order to produce a confidence interval. Assume that

$$\hat{\theta} - \theta \text{ has a distribution that, approximately, does not depend on } \theta. \quad (10.1)$$

10.1.1 The Bootstrap Sampling Scheme

Estimate $H(\hat{\theta}, \theta)$ by $H^\dagger(\hat{\theta})$, the distribution of the estimator evaluated on the set of n random vectors, $(Z_1^*, \ldots, Z_n^*)$, where Z_i^* are independent, and selected from $\{Z_1, \ldots, Z_n\}$ with probability $1/n$ for each. That is, if $\hat{\theta}_{B,i}$ are the values of $\hat{\theta}$ evaluated at each of the n^n samples, then

$$H^\dagger(\hat{\theta}) = \sum_{i=1}^{n^n} I(\hat{\theta}_{B,i} \leq \hat{\theta})/n^n.$$

Here again, the function I of a logical argument is 1 if the argument is true, and 0 if it is false. Generally, the function $\hat{\theta}$ is symmetric in its arguments (for example, $\hat{\theta}(z_1, z_2, \ldots, z_n) = \hat{\theta}(z_2, z_1, \ldots, z_n)$, and similarly for other permutations of the arguments), and so the distribution of $\hat{\theta}(Y_1, \ldots, Y_n)$ is a discrete distribution supported on fewer values than n^n, although any shortcuts that exploit this symmetry still leave the exact enumeration of distribution of $\hat{\theta}$ under the resampling distribution intractable.

Because this exhaustive approach consumes excessive resources, one almost always proceeds via random sampling. Choose a number of random samples B to draw. One draws new random samples from the population represented by the original sample, with replacement. This sampling with replacement distinguishes the bootstrap from previous permutation techniques. For random sample i, evaluate the estimator on this sample, and call it $\hat{\theta}_{B,i}$. The collection of such values is called the <u>bootstrap sample</u>. Let

$$H^*(\hat{\theta}) = \sum_{i=1}^{B} I(\hat{\theta}_{B,i} \leq \hat{\theta})/B.$$

Call this distribution the <u>resampling distribution</u>.

The Bootstrap Idea

When approximating H by H^*, two sources of errors arise: the error in approximating H by $H^\dagger$, governed by sample size, and the error in approximating $H^\dagger$ by H^*, which is governed by B. Generally speaking, moderate values for B (for example, 999, or 9999) are sufficient to make the second source of error ignorable, in the presence of the first source of data.

Techniques below will need quantiles of H^*, which are determined by ordered values of the bootstrap samples. Express the ordered values of $\hat{\theta}_{B,i}$ by $\hat{\theta}_{B,(i)}$. Order statistics from the bootstrap sample are used to estimate quantiles of the bootstrap distribution. The most naive approach uses $\hat{\theta}_{B(i)}$ to represent quantile i/B of the bootstrap distribution. By this logic, $\hat{\theta}_{B(1)}$ estimates the $1/B$ quantile, and $\hat{\theta}_{B(B)}$ represents the 1 quantile; that is, $\hat{\theta}_{B(B)}$ approximates a value with all of the true sampling distribution for $\hat{\theta}$ at or below it. Conceptually, the estimation problem ought to be symmetric if the order of bootstrap observations is swapped, but these naive quantiles are not. The upper quantile is wrong, since the population that $\hat{\theta}_B$ might be intended to represent might not exist on a bounded interval. One may make this quantile definition symmetric by taking $\hat{\theta}_{B(i)}$ to represent quantile $i/(B+1)$ of this distribution.

Bootstrap techniques below will use the analogy

$$H \text{ is to } \theta \text{ as } H^* \text{ is to } \hat{\theta}. \tag{10.2}$$

Since H^* is approximately centered about $\hat{\theta}$, bootstrap techniques will require that θ is defined so that

$$H \text{ as a function of } \hat{\theta} \text{ is centered at } \theta. \tag{10.3}$$

The statements of conditions (10.1) and (10.3) are purposely vague; Abramovitch and Singh (1985) give an early set of specific conditions, and Hall (1992) presents a manuscript-length set of tools for assessing the appropriateness of bootstrapping in various contexts. Most results guaranteeing bootstrap accuracy rely on the existence of and Edgeworth approximation to $O(1/\sqrt{n})$; that is, they require the existence of constants κ_2 and κ_3 such that (6.5) holds for the distribution of $\hat{\theta}$, with the approximate expectation κ_1 equal to θ, the terms involving κ_4 and κ_3^2 removed, and with error bounded by a constant divided by n. Such results will not apply to bootstrap inference using the sample mean for the Cauchy distribution, for example, since the variance for the Cauchy distribution is not finite. The bootstrap sometimes performs poorly when (6.5) fails to hold (Hall, 1988).

Alternatively, if one is willing to assume that F takes a parametric form, one might sample from the distribution $F(\cdot, \hat{\theta})$, and use these as above to construct H^*. This technique is called the parametric bootstrap (Efron and Tibshirani, 1993, §6.5).

10.2 Univariate Bootstrap Techniques

Various strategies exist for using these samples in the most simple univariate contexts. Terminology below is consistent with the R package boot.

10.2.1 The Normal Method

If one is willing to assume that, approximately, $\hat{\theta}$ has an approximately Gaussian distribution centered at θ and with some unknown variance, one may use the bootstrap sample to estimate the standard error of an estimator. Consider the standard error estimate

$$\check{\varsigma}^2 = \sum_{i=1}^{B}(\theta_{B,i} - \theta)^2/B. \tag{10.4}$$

Using $\bar{\theta}_B = \sum_{i=1}^{B} \theta_{B,i}/B$ in place of θ in (10.4) results in an estimate systematically too small. Use $B - 1$ instead of B in the denominator of $\check{\varsigma}^2$, or $\hat{\theta}$ in place of θ, to respond to this undercoverage. The better estimate is

$$\hat{\varsigma}^2 = \sum_{i=1}^{B}(\theta_{B,i} - \theta)^2/(B-1). \tag{10.5}$$

Then the standard deviation estimate $\hat{\varsigma}$ is the sample standard deviation of the bootstrap samples (Efron, 1981), and a $1-\alpha$ confidence interval is $\hat{\theta} \pm \hat{\varsigma} z_{1-\alpha/2}$ for $\hat{\varsigma}$ the sample standard deviation of bootstrap samples (10.5).

10.2.2 Basic Interval

Often one is unwilling to assume that the estimator is approximately Gaussian. Assume conditions (10.1) and (10.3). Use the distribution of $\hat{\theta}_{B,i} - \hat{\theta}$ as a proxy for that of $\hat{\theta} - \theta$. A confidence interval for θ may be constructed as

$$(\hat{\theta} - v_U \leq \theta \leq \hat{\theta} - v_L),$$

by determining v_L and v_U to satisfy

$$P\left[v_L \leq \hat{\theta} - \theta \leq v_U\right] = 1 - \alpha, \tag{10.6}$$

if this distribution were known.

Let u_L and u_U be $\alpha/2$ and $1 - \alpha/2$ quantiles of $\hat{\theta}_{B,i}$ respectively:

$$u_L = \theta_{B,(\alpha(B+1)/2)} \text{ and } u_U = \theta_{B,((1-\alpha/2)(B+1))}.$$

Then

$$P_*\left[u_L - \hat{\theta} \leq \theta_{B,i} - \hat{\theta} \leq u_U - \hat{\theta}\right] = 1 - \alpha, \tag{10.7}$$

Univariate Bootstrap Techniques 171

where $P_*[\cdot]$ is the probability function associated with H^*. Using analogy (10.2), equate endpoints of (10.7) and (10.6), to estimate quantiles v_U and v_L of $\hat{\theta}_{B,i} - \hat{\theta}$ by $\hat{v}_L = u_L - \hat{\theta}$, and $\hat{v}_U = u_U - \hat{\theta}$. Then a confidence interval for θ is

$$(\hat{\theta} - v_U, \hat{\theta} - v_L) = (2\hat{\theta} - \theta_{B,((1-\alpha/2)(B+1))}, 2\hat{\theta} - \theta_{B,(\alpha(B+1)/2)}).$$

This is the basic bootstrap confidence interval (Davison and Hinkley, 1997, p. 29).

10.2.3 The Percentile Method

Suppose that $\hat{\theta}$ has distribution symmetric about θ. In this case, treat $\hat{\theta} - v_U$ and $-\hat{\theta} + v_L$ as equivalent approximations to interchange, and so one can use $v_U = \hat{\theta} - u_L$, $v_L = \hat{\theta} - u_U$. The confidence interval is now

$$(u_L, u_U) = (\theta_{B,\alpha(B+1)/2}, \theta_{B,(1-\alpha/2)(B+1)})$$

(Efron, 1981).

This method is referred to as the percentile method (Efron, 1981), or as the usual method (Shao and Tu, 1995, p. 132f). If the parameter θ is transformed to a new parameter ϑ, using a monotonic transformation, then the bootstrap samples transform in the same way, and so the percentile method holds if there exists a transformation that can transform to symmetry, regardless of whether one knows and can apply this transformation.

Example 10.2.1 *Consider again the arsenic data of Example 2.3.2. We calculate a confidence interval for the median.*

```
meds<-rep(NA,999)
attach(arsenic)
for(j in seq(length(meds)))
    meds[j]<-median(sample(nails,length(nails),replace=TRUE))
```

gives the bootstrap samples. The sample *function draws a random sample with replacement. Then*

```
cat(ci<-quantile(meds,probs=c(.025,.975)),"\n")
```

gives the percentile confidence interval (0.119 0.310), and

```
cat('\n Residual Bootstrap for Median\n')
cat(ci<-2*median(nails)-rev(ci),"\n")
detach(arsenic)
```

gives the basic or residual confidence interval (0.040,0.231). Recall that the estimate of the density for the nail arsenic values was plotted in Fig-

> ure 8.1. This distribution is markedly asymmetric, and so the percentile bootstrap is not reliable; use the residual bootstrap.

10.2.4 BC_a Method

This method was introduced by Efron (1987), who called it the BC_a method. The BC_a method extends his BC method, which he terms "bias-corrected"; Efron and Tibshirani (1993) refer to the method of this section as <u>bias corrected and accelerated</u>. Again, suppose one desires a confidence interval for θ, with estimator $\hat{\theta}$. As in §10.2.3, suppose $\hat{\theta}$ can be transformed to symmetry using a transformation ϕ (which need not be known). Without loss of generality, this symmetric distribution may be taken as Gaussian. Suppose further that $\phi(\hat{\theta})$ has a standard deviation that depends linearly on $\phi(\theta)$, and that $\phi(\hat{\theta})$ has a bias that depends linearly on the standard deviation of $\phi(\hat{\theta})$. That is, assume that there exists a transformation ϕ, and constants a and ζ such that $(\phi(\hat{\theta}) - \phi(\theta))/(1 + a\phi(\theta)) + \zeta$ is approximately standard Gaussian. That is,

$$P\left[\frac{\phi(\hat{\theta}) - \phi(\theta)}{1 + a\phi(\theta)} + \zeta \leq x\right] \approx \Phi(x)$$

and

$$P\left[\phi(\hat{\theta}) \leq y\right] \approx \Phi\left(\frac{y + \zeta(1 + a\phi(\theta)) - \phi(\theta)}{1 + a\phi(\theta)}\right). \tag{10.8}$$

Let θ^* be the value of θ giving quantile $1 - \alpha$ for $\hat{\theta}$. Substituting θ^* for θ into (10.8), and equating θ with $\hat{\theta}$, gives

$$\Phi\left(\frac{-az_\alpha\zeta + \zeta(a\zeta - 2) + z_\alpha}{a(\zeta - z_\alpha) - 1}\right) = \Phi\left(\zeta + \frac{\zeta - z_\alpha}{1 - a(\zeta - z_\alpha)}\right). \tag{10.9}$$

Equating the bootstrap distribution function to this tail probability, the corresponding quantile is defined by (10.9). The quantity a is called the acceleration constant. The bias ζ may be estimated by the difference between the estimate and the median of the bootstrap samples, and a may be estimated using the skewness of the bootstrap sample.

> **Example 10.2.2** *Return again to the nail arsenic values of the previous example. We again generate a confidence interval for the median. The BC_a method, and the previous two methods, may be obtained using the package* boot.
>
> ```
> library(boot)#gives boot and boot.ci.
> #Define the function to be applied to data sets to get
> #the parameter to be bootstrapped.
> boot.ci(boot(arsenic$nails,function(x,index)
>     return(median(x[index])),9999))
> ```

Univariate Bootstrap Techniques

to give

Level	Normal	Basic
95%	(0.0158, 0.2733)	(0.0400, 0.2310)

Level	Percentile	BCa
95%	(0.119, 0.310)	(0.118, 0.277)

In this case, the normal and percentile intervals are suspect, because of the asymmetry of the distribution. The more reliable interval is the bias corrected and accelerated interval.

Recall that an exact confidence interval may be constructed using

```
library(MultNonParam); exactquantileci(arsenic$nails)
```

to obtain the interval (0.118, 0.354). Efron (1981) notes that this exact interval will generally agree closely with the percentile bootstrap approach.

One can use the bootstrap to generate intervals for more complicated statistics. The bootstrap techniques described so far, except for the percentile method, presume that parameter values over the entire real line are possible. One can account for this through transformation.

Example 10.2.3 *A similar approach may be taken to a confidence interval for the standard deviation of nail arsenic values. In this case, first change to the log scale.*

```
logscale<-function(x,index) return(log(sd(x[index])))
sdbootsamp<-boot(arsenic$nails,logscale,9999)
sdoutput<-boot.ci(sdbootsamp)
```

Figure 10.1 shows the bootstrap samples; the plot produced by

```
plot(density(sdbootsamp$t))
```

shows a highly asymmetric distribution, and the BC_a correction for asymmetry is strong. (As noted before, the actual bootstrap distribution is supported on a large but finite number of values, and is hence discrete and does not have a density; the plot is heuristic only.) The output from `boot.ci` *contains some information not generally revealed using its default printing method. In particular,* `sdoutput$bca` *is a vector with five numeric components. The first of these is the confidence level. The fourth and fifth are the resulting confidence interval end points. The second and third give quantiles resulting from (10.9). The upper quantile is very close to the maximum value of 9999;* `boot.ci` *gives a warning, and serious application of BC_a intervals would better be done using more bootstrap samples. Rerunning with*

```
boot.ci(boot(arsenic$nails,logscale,99999))$bca
```
gives the BC_a 0.95 interval (-1.647,-0.096) for the log of standard deviation of nail arsenic.

FIGURE 10.1: Boot Strap Samples for Log of Nail Arsenic Standard Deviation

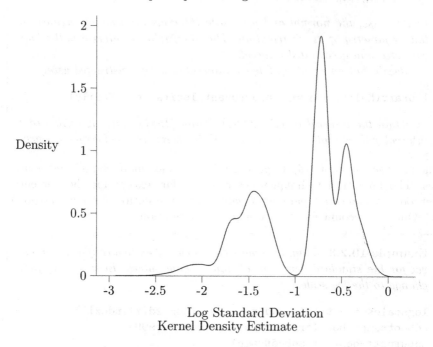

10.2.5 Summary So Far, and More Examples

Table 10.1 contains the observed coverages for nominal 0.90 confidence intervals for the medians of simulated data sets, for various distributions. Random samples of size 20 were drawn 1000 times from each distribution, and in each case, 9999 bootstrap replicates were constructed. For each random distribution, the interval was checked to see whether it contained the true population median.

TABLE 10.1: Observed coverage for nominal 0.90 Bootstrap intervals, 10 observations

	Normal	Basic	Percentile	BCa
Exponential	0.863	0.772	0.891	0.895
Uniform	0.833	0.754	0.901	0.900
Cauchy	0.917	0.835	0.875	0.862

The percentile bootstrap performed remarkably well for the exponential distribution, in light of the interval's construction assuming symmetry. None showed significant degradation when data came from a very heavy-tailed Cauchy distribution.

One can apply many of these inferential techniques to the parametric bootstrap.

Example 10.2.4 *Consider again the brain-volume data of Example 5.2.1. Treat the brain volume differences as having a Gaussian distribution. Function* boot *recognizes the parametric context through the* sim="parametric" *argument; one specifies the specific parametric assumption by providing a random number generator for bootstrap samples.*

```
cat("\nParametric Bootstrap for median difference\n")
qmed<-function(x,indices) return(median(x[indices]))
ran.diff<-function(x,ests)
   return(ests[1]+ests[2]*rnorm(length(x)))
bootout<-boot(brainpairs$diff,qmed,R=999,sim="parametric",
   ran.gen=ran.diff,mle=c(mean(brainpairs$diff),
   sd(brainpairs$diff)))
boot.ci(bootout,type=c("basic","norm"))
```

The basic interval is (-30.056, 52.720). The normal interval is almost identical.

10.3 Bootstrapping Multivariate Data Sets

The bootstrapping idea of the previous section extends directly to more complicated data contexts. The most immediate extension involves resampling data vectors as a group. When applied in regression contexts, responses per subject are often denoted by Y_j, and explanatory vector by $\boldsymbol{X}_j$. In the notation of §10.1, $\boldsymbol{Z}_j = (\boldsymbol{X}_j, Y_j)$. Consider model (9.1) and (9.2). The first approach samples the ensembles $(\boldsymbol{X}_j, Y_j)$ as a whole; that is, if a subject's response is selected for the sample, the corresponding explanatory variables will be also selected, and with the same multiplicity. This is called a random X bootstrap.

Example 10.3.1 *Consider again the brain-volume data of Example 5.2.1. We use a random X bootstrap to get confidence intervals for the*

Pearson correlation between first and second twin brain volumes. First, define the correlation function

```
rho<-function(d,idx) return(cor(d[idx,1],d[idx,2]))
```

and calculate the bootstrap samples

```
bootout<-boot(brainpairs[,c("v1","v2")],rho,9999)
```

for pairs of data. Figure 10.2 presents a histogram of the sample, with vertical lines marking the estimate and percentile confidence interval end points:

```
hist(bootout$t,freq=FALSE)
legend(-.4,6,lty=1:2,
    legend=c("Estimate","Confidence Bounds"))
abline(v=bootout$t0)
abline(v=sort(bootout$t)[(bootout$R+1)*c(.025,.975)],lty=2)
```

Confidence intervals may be calculated using

```
boot.ci(bootout)
```

to give

Level	Normal	Basic
95%	(0.7463, 1.1092)	(0.8516, 1.1571)

Level	Percentile	BCa
95%	(0.6718, 0.9773)	(0.5609, 0.9719)

To repeat, the percentile interval is represented in Figure 10.2 by broken lines. The basic (that is, residual) interval is the percentile interval reflected about the estimate. This forces the upper end point above 1, and is outside allowable values for a correlation coefficient. Similarly, the variance used for the normal interval is increased by the long tail to the left, pushing the upper bound above 1. The BC_a approach is tailored to the asymmetry in the bootstrap replicates, and is more reliable.

10.3.1 Regression Models and the Studentized Bootstrap Method

We apply random X bootstrap techniques to inference on a regression parameter. The ordinary least-squares regression estimator has a standard error that depends on the covariate patterns in the data set, and hence regression parameters from random bootstrap samples from a data set will have different precisions associated with them. Hence members of a collection of bootstrapped regression parameters are not identically distributed.

Bootstrapping Multivariate Data Sets 177

FIGURE 10.2: Histogram of Bootstrap Samples of Brain Volume Correlations

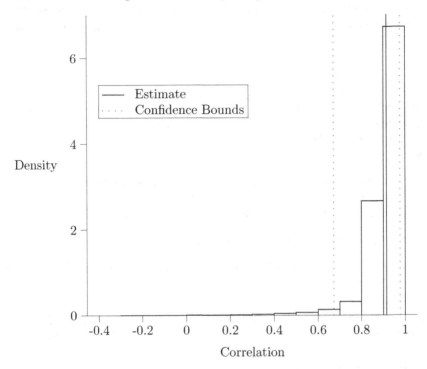

Recall the Gaussian-theory approach to constructing confidence intervals for a regression parameter $\hat{\theta} \pm t_{1-\alpha/2} \times \text{s.e.}(\hat{\theta})$, where t is a $\mathcal{T}$ quantile, with an appropriate degree of freedom. The ratio $(\hat{\theta} - \theta)/\text{s.e.}(\hat{\theta})$ is called <u>Studentized</u>, in that it is divided by an empirical estimator of its variance in order to make it comparable to a reference distribution. This section applies this idea to bootstrap sampling.

Suppose that bootstrap samples are still independent, but, as in the previous section, the variance of the estimator may depend on the unknown parameter. Suppose furthermore that a practical estimate of the variability of the estimator, on the scale of the original data, also exists. Then, re-sample B data sets as above, and for each re-sampled data set, calculate the estimate $\hat{\theta}_{B,i}$, and an estimate of variability $\hat{\sigma}_{B,i}$. Let empirical distribution of $T_{B,i} = (\hat{\theta}_{B,i} - \hat{\theta})/\hat{\sigma}_{B,I}$ stand in for distribution of $T = (\hat{\theta} - \theta)/\sigma$. This distribution doesn't involve unknown parameters; such a quantity is called <u>pivotal</u>. One might apply this in conjunction with the percentile method, as in §10.2.3. In order to construct a $1 - \alpha$ confidence interval, let t_L and t_U be the $\alpha/2$ and $1 - \alpha/2$ quantiles from bootstrap distribution.

Hall (1992, §1.3) notes that, unlike some of the other bootstrapping methods, the Studentized bootstrap method is not invariant to increasing non-affine transformations of the parameter; that is, for example, Studentized bootstrap confidence intervals for $\exp(\theta)$ are not the exponentiation of Studentized bootstrap confidence intervals for θ, and, furthermore, the quality of the Studentized bootstrap confidence intervals depends on the effectiveness of Studentization.

> **Example 10.3.2** *Consider the Studentized interval applied to the brain-volume data of Example 5.2.1. The biggest innovation here is to add standard error information to the bootstrap samples. In R, this is done by having the function to be run on the data samples return a vector with two components. The second component is the measure of variability, and should be on the squared scale.*
>
> ```
> #Second component is an estimate of its scale, squared.
> regparam<-function(d,idx) return(summary(
>     lm(d[idx,2]~d[idx,1]))$coefficients[2,1:2]^(1:2))
> ```
>
> *Now the calculations may be run as before:*
>
> ```
> boot.ci(boot(brainpairs[,c("v1","v2")],regparam,9999))
> ```
>
> *to obtain*
>
> ```
> Level      Normal              Basic              Studentized
> 95%   (0.2836, 1.1388)   ( 0.1746,  0.9239)   (0.4733, 1.0067)
>
> Level      Percentile          BCa
> 95%   (0.6296, 1.3788)   ( 0.5962, 1.2082)
> ```
>
> *Note that the warning about missing information for the Studentized interval no longer appears. The appropriate interval is the Studentized interval; use that one.*

10.3.2 Fixed X Bootstrap

In certain contexts, resampling explanatory and response variables together may be unrealistic. For example, one might be analyzing a designed experiment, with a certain number of responses specified in advance for each of a set number of explanatory variable patterns. In this case, one might want to keep the observed pattern of explanatory variables, and associate with them resampled responses. A similar idea was used in permutation testing earlier. The present approach differs from the permutation approach in at least two ways. Most trivially, the bootstrap approach involves resampling with replacement, and the permutation approach involves resampling without replacement. More fundamentally, the sampling distribution for permutation testing needs to hold only for the null distribution; in contrast, the bootstrap approach should be constructed to produce a pivotal distribution independent of the value of the

Bootstrapping Multivariate Data Sets

regression parameter. The next example exhibits resampling residuals; some refinements and a second example follow that produce a technique more closely following the regularity conditions for the bootstrap.

Example 10.3.3 *Refer again to the brain size data of Example 5.2.1. In order to apply the fixed-X bootstrap, first calculate fitted values and residuals:*

```
residuals<-resid(lm(brainpairs$v2~brainpairs$v1))
fittedv<-brainpairs$v2-residuals
```

If a function needs a variable that is not locally defined, R looks for it to be defined globally. Using the function **boot** *inside another function, and manipulating the data on a level above the bootstrap function but below the command line, may cause the code to fail, or give something other than what was intended. In the example below,* **fittedv** *and* **residuals** *are passed explicitly to* **boot**, *after the data, indices, and bootstrap sample size, and must be referred to by name.*

```
regparam<-function(data,indices,fittedv,residuals){
  y<-fittedv+residuals[indices]
  return(summary(lm(y~data[,1]))$coefficients[2,1:2]^(1:2))
}
library(boot)
bootout<-boot(brainpairs[,c("v1","v2")],regparam,9999,
   fittedv=fittedv,residuals=residuals)
```

Again, the Studentized interval is appropriate:

```
boot.ci(bootout,type="stud")
```

gives the 0.95 interval (0.5100, 1.0509).

Without a hypothesis of a zero regression parameter, residuals remain with expectation zero, but have a variance that depends on the distance of the explanatory vector from the center of the set of explanatory vectors. Hence when performing inference on regression parameters by resampling residuals, the quantities re-sampled are not independent and identically distributed. The variance matrix of the residuals is $\sigma^2(I - H)$, for X is the matrix of covariate vectors, as defined in (9.4), and for $H = X(X^\top X)X^\top$ (Hoaglin and Welsch, 1978). One might bring residuals closer to the identically-distributed case by first dividing the residuals by $1 - h_i$, for h_i the diagonal elements of H, resampling these rescaled residuals, and then multiplying the re-sampled residual by $1 - h_i$. The values of h_i are specific to the covariate pattern, and will generally be applied to residuals arising from different observations when contributing to the divisor and the multiplier.

Example 10.3.4 *Refer again to the brain size data of Example 5.2.1. Modify the approach of Example 10.3.3:*

```
cat("\n Fixed X bootstrap adjusting for variance   \n")
modelfit<-lm(brainpairs$v2~brainpairs$v1)
hat<-lm.influence(modelfit)$hat
adjresid<-resid(modelfit)/sqrt(1-hat)
newregparam<-function(data,indices,fittedv,adjresid,hat){
  y<-fittedv+sqrt(1-hat)*adjresid[indices]
  return(summary(lm(y~data[,1]))$coefficients[2,1:2]^(1:2))
}
bootout<-boot(brainpairs[,c("v1","v2")],newregparam,9999,
    fittedv=fittedv,adjresid=adjresid,hat=hat)
boot.ci(bootout,type="student")
```

gives the 0.95 Studentized interval (0.5654,0.9951). Proper adjustment for differences in variance results in a tighter confidence interval.

In this case, because the covariate pattern is fixed, the standard error of the regression coefficient varies between bootstrap samples only because of differences in the estimated residual standard deviation, and so the case for the Studentized interval is less compelling. The BC_a interval, (0.6030,0.9416), obtainable by changing the argument to `boot.ci`*, is appropriate in this case.*

The above example demonstrates how to adjust for unequal variances for residuals; adjusting for a lack of independence is more difficult.

One might bootstrap the location difference between two data sets. This two-sample location model is a nonparametric version of the classical two-sample t confidence interval setting. It may be mimicked using a linear models approach (9.1) and (9.2) by using a single covariate vector, taking the value 0 for one group and 1 for the other group, to give the pooled version of t intervals; the slope parameter is then the difference in locations for the two groups, and fitted values are group means. In this case, the fixed-X approach compares the mean difference estimate to other samples with the same number of observations in the first group as is observed in the data set, with a similar statement about the second group, and the random-X approach fails to do this. In this respect, the fixed-X approach is more intuitive.

The same approach may be applied to explore differences among a larger number of groups. One might use the bootstrap to give confidence bounds for a wide variety of possible ANOVA summaries.

Example 10.3.5 *Refer again to the chicken weight gain data of Example 5.4.2. Use fixed-X bootstrap techniques to bound the largest group mean minus the smallest group mean, when data are split by protein level.*

The Jackknife

```
meandiff<-function(data,indices){
  fitv<-fitted(lm(data[,1]~as.factor(data[,2])))
  residv<-data[,1]-fitv
  y<-fitv+residv[indices]
  return(diff(range(unlist(lapply(split(y,data[,2]),
    "mean")))))
}
attach(chicken)
cat("Bootstrap CI for maximal difference in means\n")
boot.ci(boot(cbind(weight,lev),meandiff,9999))
```

Because the design is balanced, with an equal number of chickens at each level of diet, no adjustment for different standard deviations of the residuals is needed. Because the fixed-X bootstrap is employed, Studentization of the resulting mean differences is not necessary. The resulting BC_a confidence interval is (85.6, 769.5).

A similar approach might have been employed for inference about quantities like the R^2.

10.4 The Jackknife

Consider a related technique, the jackknife (Quenouille, 1956; Miller, 1974), a technique to estimate the bias of an estimator, using values of the estimator evaluated on subsets of the observed sample, under certain conditions on the form of the bias. Suppose that the bias of estimator T_n of θ based on n observations is

$$\mathrm{E}\left[T_n\right] - \theta = a/n + O(n^{-3/2}), \tag{10.10}$$

for some unknown a. Here $O(n^{-3/2})$ denotes a quantity which, after multiplication by $n^{3/2}$, is bounded for all n,

10.4.1 Examples of Biases of the Proper Order

Quenouille (1956) suggested this technique in situations where (10.10) held with at least one more term of form b/n^2, and with a correspondingly smaller error. That is, (10.10) is replaced with

$$\mathrm{E}\left[T_n\right] - \theta = a/n + b/n^2 + O(n^{-5/2}). \tag{10.11}$$

For example, if $X_1, \ldots, X_n$ are independent, and identically distributed with distribution $\Phi(\mu, \sigma^2)$, the maximum likelihood estimator for σ^2 is $\hat{\sigma}^2 = \sum_{j=1}^n (X_j - \bar{X})^2/n$. Note that $\tilde{\sigma}^2 = n\hat{\sigma}^2/(n-1)$ is an unbiased estimator of σ^2, since $\mathrm{E}_{\sigma^2}\left[\hat{\sigma}^2\right] = (n-1)\sigma^2/n = \sigma^2 - \sigma^2/n$. Then (10.10) holds with $a = \sigma^2$ and $b = 0$.

Restriction (10.11) seems unnecessarily strict, and cases in which it holds are less common. For example, let W_n be an unbiased estimate of a parameter ω, and let $T_n = g(W_n)$, for some smooth function g. Let $\theta = g(\omega)$, and assume that $\text{Var}[W_n] \approx a/n$. Then

$$\begin{aligned} g(W_n) &\approx g(\omega) + g'(\omega)(W_n - \omega) + g''(\omega)(W_n - \omega)^2/2 \\ &\quad + g'''(\omega)(W_n - \omega)^3/6 + g''''(\omega)(W_n - \omega)^4/24, \end{aligned}$$

and

$$\begin{aligned} \mathrm{E}[g(W_n)] &\approx g(\omega) + g''(\omega)\text{Var}[W_n]/2 \\ &\quad + g'''(\omega)\mathrm{E}\left[(W_n - \omega)^3\right]/6 + g''''(\omega)\mathrm{E}\left[(W_n - \omega)^4\right]/24. \end{aligned}$$

Then (10.10) holds, and (10.11) holds if the skewness of W_n times $\sqrt{n}$ converges to 0.

10.4.2 Bias Correction

Let $T^*_{n-1,i}$ be the estimator based on the sample of size $n-1$ with observation i omitted. Let $\bar{T}^*_n = \sum_{i=1}^{n} T^*_{n-1,i}/n$. Then the bias of $\bar{T}^*_n$ is approximately $a/(n-1)$. Under (10.10), $B = (n-1)(\bar{T}^*_n - T_n)$ estimates the bias, since

$$\begin{aligned} \mathrm{E}[B] &= (n-1)(\theta + a/(n-1) - \theta - a/n) + O(n^{-3/2}) \\ &= a/n + O(n^{-3/2}). \end{aligned}$$

Then the bias of $T_n - B$ is $O(n^{-3/2})$. Furthermore, under the more stringent requirement (10.11), the bias is $O(n^{-2})$.

10.4.2.1 Correcting the Bias in Mean Estimators

Sample means are always unbiased estimators of the population expectation. Consider application of the jackknife to the sample mean. In this case, $T_n = \sum_{j=1}^{n} X_j/n$, $T^*_{n-1,i} = \sum_{j=1, j \neq i}^{n} X_j/(n-1)$, and

$$\bar{T}^*_{n-1} = \sum_{i=1}^{n} \sum_{j=1, j \neq i}^{n} X_j/(n(n-1)) = \bar{X},$$

and hence the bias estimate is zero.

10.4.2.2 Correcting the Bias in Quantile Estimators

Consider the jackknife bias estimate for the median for continuous data, and, for the sake of defining the $T^*_{n-1,i}$, let i index the order statistic $X_{(i)}$.

When the sample size n is even, then $T_n = (X_{(n/2)} + X_{(n/2+1)})/2$, and

$$T^*_{n-1,i} = \begin{cases} X_{(n/2+1)} & \text{if } i \leq n/2 \\ X_{(n/2)} & \text{if } i \geq n/2 + 1. \end{cases}$$

The Jackknife

Then $\bar{T}^*_{n-1} = (X_{(n/2)} + X_{(n/2+1)})/2 = T_n$, and the bias estimate is always 0. For n odd, then $T_n = X_{((n+1)/2)}$, and

$$T^*_{n-1,i} = \begin{cases} (X_{((n-1)/2)} + X_{((n+1)/2)})/2 & \text{if } i > (n+1)/2 \\ (X_{((n+3)/2)} + X_{((n+1)/2)})/2 & \text{if } i < (n+1)/2 \\ (X_{((n+3)/2)} + X_{((n-1)/2)})/2 & \text{if } i = (n+1)/2, \end{cases}$$

and the average of results from the smaller sample is

$$\bar{T}^*_{n-1} = \tfrac{n+1}{4n} X_{((n-1)/2)} + \tfrac{n+1}{4n} X_{((n+3)/2)} + \tfrac{n-1}{2n} X_{((n+1)/2)}.$$

Hence

$$\begin{aligned}\bar{T}^*_{n-1} - T_n &= \left(\tfrac{n+1}{4n} X_{((n-1)/2)} + \tfrac{n+1}{4n} X_{((n+3)/2)} + \tfrac{n-1}{2n} X_{((n+1)/2)}\right) - X_{((n+1)/2)} \\ &= \tfrac{n+1}{4n}\left(X_{((n-1)/2)} + X_{((n+3)/2)} - 2X_{((n+1)/2)}\right), \quad (10.12)\end{aligned}$$

and the bias estimate is

$$B = (n-1)(\bar{T}^*_{n-1} - T_n) = \tfrac{n^2-1}{4n}\left(X_{((n-1)/2)} + X_{((n+3)/2)} - 2X_{((n+1)/2)}\right).$$

Example 10.4.1 *Consider again the nail arsenic data of Example 2.3.2. Calculate the Jackknife estimate of bias for the mean, median, and trimmed mean for these data. Jackknifing is done using the* `bootstrap` *library, using the function* `jackknife`.

```
library(bootstrap)#gives jackknife
jackknife(arsenic$nails,median)
```

This function produces the 21 values, each with one observation omitted:

```
$jack.values
 [1] 0.2220 0.2220 0.2220 0.2220 0.1665 0.1665 0.2220 0.2220
 [9] 0.1665 0.2220 0.2220 0.1665 0.1665 0.1665 0.1665 0.1665
[17] 0.1665 0.2220 0.1665 0.2220 0.2135
```

Each of these values is the median of 20 observations. Ten of them, corresponding to the omission of the lowest ten values, are the averages of $X_{(10)}$ and $X_{(11)}$. Ten of them, corresponding to the omission of the highest ten values, are the averages of $X_{(9)}$ and $X_{(10)}$. The last, corresponding to the omission of the middle value, is the average of $X_{(9)}$ and $X_{(11)}$. The mean of the jackknife observations is 0.1952. The sample median is 0.1750, and the bias adjustment is $20 \times (0.1952 - 0.1750) = 20 \times 0.0202 = 0.404$, as is given by R:

```
$jack.bias
```

[1] 0.4033333

This bias estimate for the median seems remarkably large. From, (10.12) the jackknife bias estimate for the median is governed by the difference between the middle value and the average of its neighbors. This data set features an unusually large gap between the middle observation and the one above it.

Applying the jackknife to the mean via

```
jackknife(arsenic$nails,mean)
```

gives the bias correction 0:

```
$jack.bias
[1] 0
```

as predicted above. One can also jackknife the trimmed mean:

```
jackknife(arsenic$nails,mean,trim=0.25)
```

The above mean is defined as the mean of the middle half of the data, with 0.25 proportion trimmed from each end. Although the conventional mean is unbiased, this unbiasedness does not extend to the trimmed mean:

```
$jack.bias
[1] -0.02450216
```

In contrast to the arsenic example with an odd number of data points, consider applying the jackknife to the median of the ten brain volume differences, from Example 5.2.1:

```
attach(brainpairs);jackknife(diff,median)
```

to give 0 as the bias correction.

The average effects of such corrections may be investigated via simulation. Table 10.2 contains the results of a simulation based on 100,000 random samples for data sets of size 11. In this exponential case, the jackknife bias correction over-corrects the median, but appears to address the trimmed mean exactly.

TABLE 10.2: Expectations of statistic and Jackknife bias estimate

Distribution	Statistic T	E$[T]$	E$[B]$	Parameter
Exponential	mean	0.998	0.000	1
Exponential	median	0.738	0.091	0.693
Exponential	0.25 trimmed mean	0.810	-0.060	0.738

Under some more restrictive conditions, one can also use this idea to estimate the variance of T.

10.5 Exercises

1. The data set

    http://ftp.uni-bayreuth.de/math/statlib/datasets/lupus

    gives data on 87 lupus patients. The fourth column gives transformed disease duration.

    a. Give a 90% bootstrap confidence interval for the mean transformed disease duration, using the basic, Studentized, and BCa approaches.

    b. Give a jackknife estimate of the bias of the mean and of the 0.25 trimmed mean transformed disease duration (that is, the sample average of the middle half of the transformed disease duration).

2. The data set

    http://ftp.uni-bayreuth.de/math/statlib/datasets/federalistpapers.txt

    gives data from an analysis of a series of documents. The first column gives document number, the second gives the name of a text file, the third gives a group to which the text is assigned, the fourth represents a measure of the use of first person in the text, and the fifth presents a measure of inner thinking. There are other columns that you can ignore. (The version at Statlib, above, has odd line breaks. A reformatted version can be found at

    stat.rutgers.edu/home/kolassa/Data/federalistpapers.txt).

    a. Calculate a bootstrap confidence interval, with confidence level .95, for the regression coefficient of inner thinking regressed on first person. Test at $\alpha = .05$. Provide basic, Studentized, and BCa intervals. Do the fixed-X bootstrap.

    b. Calculate a bootstrap confidence interval, with confidence level .95, for the regression coefficient of inner thinking regressed on first person. Provide basic, Studentized, and BCa intervals. Do not do the fixed-X bootstrap; re-sample pairs of data.

    c. Calculate a bootstrap confidence interval, with confidence level .95, for the R^2 statistic for inner thinking regressed on first person. Provide basic and BCa intervals. Do not do the fixed-X bootstrap; re-sample pairs of data.

A
Analysis Using the SAS System

The preceding chapters detailed statistical computations using R. This appendix describes parallel computations using SAS. Some of the computations require the use of macros, available at

http://stat.rutgers.edu/home/kolassa/Data/common.sas .

Include this file before running any macros below, by first downloading into the current working directory, and adding

%include "/folders/myfolders/common.sas";

to your SAS program before using the macros. Adjust the local folder /folders/myfolders/ to reflect your local configuration. Furthermore, the examples use data sets that need to be read into SAS before analysis; reading the data set is given below the first time the data set is used.

Example 2.3.2: Perform the sign test to evaluate the null hypothesis that the population median for nail arsenic levels is .26. This is done using proc univariate:

```
/***************************************************************/
/* Data from http://lib.stat.cmu.edu/datasets/Arsenic           */
/* reformatted into ASCII on the course home page.  Data re-    */
/* flect arsenic levels in toenail clippings; covariates in-    */
/* clude age, sex (1=M), categorical measures of quantities     */
/* used for drinking and cooking, arsenic in the water, and     */
/* arsenic in the nails.  To make arsenic.dat from Arsenic, do*/
/*antiword Arsenic|awk '((NR>39)&&(NR<61)){print}'>arsenic.dat*/
/* Potential threshold for ill health effects for toenails is  */
/* .26 http://www.efsa.europa.eu/de/scdocs/doc/1351.pdf         */
/***************************************************************/
data arsenic; infile '/folders/myfolders/arsenic.dat';
    input age sex drink cook water nails; run;
proc univariate data=arsenic mu0=.26 ciquantdf(alpha=.05);
    var nails; run;
```

Material between /* and */ are ignored by SAS, and are presented above so that information describing the data set may be included with the analysis code. The option mu0=.26 specifies the null hypothesis, and ciquantdf(alpha=.05) specifies that distribution free intervals for quantiles, with confidence .95=1-.05, should be produced. The var statement indicates which variable should be summarized.

The hypothesis test might also have been done using proc freq:

```
data arsenic; set arsenic;y=nails>.26; run;
proc freq data=arsenic; tables y/binomial(p=.5 level=1);
   exact binomial; run;
```

Example 2.3.3: The proc univariate call above gives intervals for a variety of quantiles, including quartiles, but not for arbitrary user-selected quantiles. The following code gives arbitrary quantiles. Manually edit the code below to replace 21 with the actual number of observations, 0.025 with half the desired complement of the confidence, and .75 with the desired quantile.

```
proc sort data=arsenic; by nails; run;
data ci; set arsenic;
   a=quantile("binomial",.025,.75,21);
   b=21+1-quantile("binomial",.025,1-.75,21);
   if _N_=a then output; if _N_=b then output; run;
title 'Upper quartile .95 CIs for nail arsenic';
proc print data=ci; run;
```

Example 2.5.1: Calculations for the empirical cumulative distribution function may be performed using

```
proc univariate data=arsenic; var nails; cdfplot nails; run;
```

and graphical output may be adjusted to your local SAS installation as necessary.

Example 3.3.1: This example calculates Mood's median test for the subset of yarn strength data coming from bobbin 3. Note that the variable denoting type is a character variable; this is designated by $. First, construct the reduced data set. Then, use proc npar1way to run the test. This procedure does the bulk of nonparametric group comparison calculations available in SAS; two-group comparisons are a special case. Group membership is specified by the variable specified in the class statement. The median keyword triggers Mood's median test.

```
/**********************************************************/
/* Yarn strength data from Example Q of Cox & Snell      */
/* (1981).  Variables represent strength of two types of*/
/* yarn collected from six different bobbins.            */
/**********************************************************/
data yarn; infile '/folders/myfolders/yarn.dat';
   input strength bobbin type $; run;
data yarn1; set yarn;
```

Analysis Using the SAS System

```
         if bobbin=3 then;   else delete;  run;
title 'Median test applied to bobbin 3 yarn data';
proc npar1way data=yarn1 median ; class type;
   exact; var strength; run;
```

Compare this with the *t*-test results.

```
proc ttest data=yarn1 ; class type; var strength; run;
```

Example 3.4.1: This example applied the Wilcoxon rank sum test to investigating differences between the strengths of yarn types. At present, analysis is restricted to bobbin 3, since this data set did not have ties. These score tests are calculated using `proc npar1way`, with the option `wilcoxon`. The first has continuity correction turned off, with the option `correct=no`. Correction is on by default, and so no such option appears in the second run, with correction. Exact values are given using the `exact` statement, followed by the test whose exact *p*-values are to be calculated.

```
title 'Wilcoxon Rank Sum Test, yarn strength by type, bobbin 3';
title2 'Approximate values, no continuity correction';
proc npar1way data=yarn1 wilcoxon correct=no ;
   class type; var strength; run;
title2 'Approximate values, continuity correction';
proc npar1way data=yarn1 wilcoxon ; class type;
   var strength; run;
title2 'Exact values';
proc npar1way data=yarn1 wilcoxon ; class type;
   exact wilcoxon; var strength; run;
```

Example 3.4.2: Scores are requested as part of `proc npar1way` statement.

```
proc npar1way data=arsenic vw savage ;
   class sex; var nails; run;
```

Example 3.4.3: Permutation testing may be done using `proc npar1way` and scores=data:

```
title 'Permutation test for Arsenic Data';
proc npar1way data=arsenic scores=data ;
   class sex; exact scores=data; var nails; run;
```

Tables 3.4 and 3.5: Test sizes and powers may be calculated using the macro `test2` in the file `common.sas`, as above. First and second arguments are the group sizes. The third argument is the Monte Carlo sample size. The last argument is the offset between groups.

```
title 'Monte Carlo Assessment of Test Sizes';
%include "/folders/myfolders/common.sas";
```

```
%test2(10,10,10000,0); proc print data=size noobs; run;
title 'Monte Carlo Assessment of Test Powers';
%test2(10,10,10000,1); proc print data=size noobs; run;
```

Example 3.9.1: The npar1way procedure also performs the Siegel-Tukey and Ansari-Bradley tests, using the keywords ab and st in the statement.

```
proc npar1way data=yarn ab st ; class type; var strength; run;
```

Example 3.10.1: Hodges-Lehmann estimation is performed by npar1way, using the hl option to the proc npar1way statement. Adding the exact hl first inverts the Mann-Whitney-Wilcoxon test to determine which order statistics make up the interval; otherwise, a Gaussian approximation to the Mann-Whitney-Wilcoxon test is used to get critical values, and confidence interval end points are interpolated between order statistics.

```
proc npar1way data=arsenic hl wilcoxon plots=none;
   class sex; exact hl ; var nails; run;
```

Example 3.11.1: Kolmogorov-Smirnov and Cramér-von Mises tests are also performed by proc npar1way, by specifying edf in the statement:

```
proc npar1way data=yarn edf; class type; var strength; run;
```

Examples 4.3.1 and 4.4.1: The Kruskal-Wallis test, and the variants using the Savage and the van der Waerden scores, can be performed using proc npar1way.

```
data maize; infile '/folders/myfolders/T58.1' ;
   input exno tabno lineno loc $ block $ plot
      treat $ ears weight;
   nitrogen=substr(treat,1,1); if weight<0 then weight=.; run;
data tean; set maize; if loc="TEAN" then; else delete;
   if weight=. then delete; run;
title 'Kruskal Wallis H Test for Maize Data';
proc npar1way wilcoxon savage vw data=tean plots=none;
   class treat; var weight; /* exact;*/ run;
```

The option exact is commented out above; depending on the configuration of the example, exact computation can be very slow, and resource-intensive enough to make the calculations fail. The option wilcoxon triggers the Kruskal-Wallis test, since it uses Wilcoxon scores. These calculations may be compared with the standard Analysis of Variance approach.

```
proc anova data=tean; class treat; model weight=treat; run;
```

Example 4.4.2: Here are two calls to npar1way that give in principle the same answer. The exact command causes exact *p*-values to be computed. If you

Analysis Using the SAS System

specify scores (in this case data) in the exact statement, you'll get that exact p-value. If you don't specify scores, you'll get the exact p-value for the scores specified in the `proc nparlway` statement. If you don't specify either, you'll get ranks without scoring. As we saw before, exact computations are hard enough that SAS quits. The second `proc nparlway` presents a compromise: random samples from the exact distribution. Give the sample size you want after `/mc`. The two calls give different answers, since one is approximate. Successive calls will give slightly different answers.

```
title 'Permutation test for Maize Data';
proc nparlway data=tean scores=data plots=none; class treat;
    exact ; var weight; run;
title1 'Maize Permutation test, Monte Carlo Version';
proc nparlway data=tean scores=data plots=none; class treat;
    exact scores=data/mc n=100000 ; var weight; run;
```

Example 4.7.5: Here we test for an ordered effect of nitrogen, and compare with the unordered version.

```
title 'Jonckheere-Terpstra Test for Maize';
proc freq data=tean noprint; tables weight*nitrogen/jt;
    output out=jttab jt; run;
proc print data=jttab noobs; run;
title 'K-W Test for Maize, to compare with JT';
proc nparlway data=tean plots=none; class nitrogen;
    var weight; run;
```

Example 5.2.1: Perform a paired test on brain volume, assuming symmetry. First, read the data:

```
data twinbrain; infile 'twinbrain.dat';
    input CCMIDSA FIQ HC ORDER PAIR SEX TOTSA TOTVOL WEIGHT; run;
data fir; set twinbrain; v1=totvol; if order=1 then output; run;
data sec; set twinbrain; v2=totvol; if order=2 then output; run;
data brainpair; merge fir sec; by pair; diff=v2-v1; run;

title '1-sample sign, T, and Wilcoxon tests for brain volume';
proc univariate data=brainpair; var diff; run;
```

Examples 5.4.1 and 5.6.1: Perform a paired test on brain volume, assuming symmetry. First, read the data:

```
data expensesd; infile '/folders/myfolders/friedman.dat';
    input cat $ g1 g2 g3 g4 g5 g6 g7; run;
```

Convert to a set with one observation per line, with level 1 as well:

```
data friedman; set expensesd; cate=_n_;
   l=1; sd=g1; output; l=2; sd=g2; output; l=3; sd=g3; output;
   l=4; sd=g4; output; l=5; sd=g5; output; l=6; sd=g6; output;
   l=7; sd=g7; output; run;
```

Now calculate the test statistics:

```
proc freq data=friedman;
   table cate*l*sd/cmh2 noprint score=rank; run;
```

Friedman's test is the six degree of freedom test. Page's test, in the non-replicate balanced case, coincides with the one-degree of freedom correlation test above.

Example 6.3.1: This code performs the permutation test. All intermediate values are saved, resulting in an inefficient use of computer memory.

```
data big; set brainpair;
     do i=1 to 200000 ; u=rand("uniform"); output; end; run;
* Here we have to sort three times.  When big is created,;
* set contains 200000 lines for twin 1, then 200000 for;
* twin 2, etc.  We will need to rank observations by us, within;
* each simulated data set, and proc rank needs these sorted;
* by i.  Then we need to associate the new ranks of second obs.;
* with the old first obs. by merging by subject, and so we sort;
* data by subject.  Finally, we need to calculate correlations;
* separately for each simulated data set, and so we sort by i.;
* There has to be a more efficient way to do this, but I can;
* not figure it out.;
proc sort data=big; by i; run;
proc rank data=big out=big; var u; by i; ranks pair; run;
proc sort data=big; by pair; run;
data big; merge big (drop=v1) brainpair (drop=v2); by pair; run;
proc sort data=big; by i; run;
proc reg data=big outest=rgout noprint; model v2=v1; by i; run;
proc reg data=brainpair outest=oneout noprint; model v2=v1; run;
data oneout; set oneout ; v10=v1; keep v10 _type_; run;
data rgout ; merge oneout rgout; by _type_; a=1;
     if v1<v10 then a=0; run;
proc means data=rgout; var a; run;
```

The permutation *p*-value is produced as the means in the last call above.

Example 6.3.1: Only the parametric analysis is shown below.

```
title 'One-sample Multivariate analyses';
* The h in manova statement below is syntax and NOT a variable
```

Analysis Using the SAS System

```
* in the data set.  Exact does NOT do a permutation test.;
proc glm data=bp; model spd dpd =;
   manova h=INTERCEPT/mstat=exact; run;
```

Example 7.5.1: Calculate the parametric analysis for comparison purposes:

```
title 'One-sample Multivariate analyses';
* The h in manova statement below is syntax and NOT a variable
* in the data set.  Exact does NOT do a permutation test.;
proc glm data=bp; model spd dpd =;
   manova h=INTERCEPT/mstat=exact; run;
```

Brute-force permutation testing of the wheat data may be performed using

```
proc sort data=big; by i; run;
proc rank data=big out=big; var u; by i; ranks item; run;
proc sort data=big; by item; run;
data big; merge big (drop=region) wheat; by item; run;
proc sort data=big; by i; run;
* Block nearly identical to the permutation test for;
* correlation ends.;
* a and h following the model statement are variable names;
* and are short for Atou and Huntsman. h following manova;
* is part of the manova syntax and will be the same regard-;
* less of variable names.;
proc glm data=big; class region; by i; model a h=region;
   ODS output MultStat=a;
   manova h=region/mstat=exact; run;
proc ttest data=big plots=none;
   ODS output TTests=ttestout;
   class region; by i; var a h; run;
data ttestout; set ttestout; keep i Variable tValue;
   if method='Satterthwaite' then delete; run;
proc means data=ttestout max noprint; by i;var tValue;
   output out=ttmax max=tValue; run;
* Data set a has four lines per permutation.  We can keep any;
* of these.  It does not matter which.;
data a; set a; by i; if first.i then; else delete; run;
data both; merge a ttmax; by i;
   keep FValue tValue i Hypothesis; run;
data obss; set obss; TObs=tValue;FObs=FValue;
   keep TObs FObs Hypothesis; run;
* Merge by any variable that is constant in testout.  Proc glm;
* with manova creates a variable Hypothesis that does this.;
data both ; merge both obss; by Hypothesis;
   countf=0; if FValue ge FObs then countf=1;
   countt=0; if tValue ge TObs then countt=1;  run;
proc means data=both mean; var countt countf; run;
```

Example 8.2.1: SAS does density estimation via proc univariate. Only graphical output is required, and so **noprint** is specified. The histogram is plotted by default.

```
proc univariate data=arsenic noprint;
    histogram nails/kernel (k=normal )
    kernel(k=triangular )
    kernel(k=quadratic ) ; run;
title 'Kernel density est. with an excessively small bandwidth';
proc univariate data=arsenic noprint;
    histogram nails/kernel (k=normal color=blue c=.001) ; run;
```

The following is another way to draw density estimates. This method allows only the normal kernel. Syntax has changed with versions of SAS, and so this will run only with recent versions. There is no reason to see the text output, but suppressing text output also suppresses graphics. There is no overall **noprint** option.

```
title1 'Kernel smoother for arsenic via kde';
proc kde data=arsenic ; univar nails/plots=density; run;
```

Example 9.2.2: The default degree in loess in SAS is 1.

```
proc loess data=arsenic plots(only)=fit;
    model nails=water/degree=2 all; run;
proc loess data=both plots(only)=fit; model v2=v1; run;
```

Example 9.3.1: Various sources suggest that isotonic regression can be done with the macro at

> http://www.bios.unc.edu/distrib/redmon/SASv8/redmon2.sas .

Include the macro using

```
%include "redmon2.sas"
```

I was not able to make this work.

Example 9.4.1: Fit the spline using

```
proc transreg data=arsenic ;
    model identity(nails)=spline(water); run;
```

The option **noprint** suppresses graphical output, and so it is not used.

Example 9.5.1: Perform these calculations using

```
proc quantreg data=arsenic; model nails=water;   run;
```

Analysis Using the SAS System

Example 9.5.2: Fit the median and 0.2 quantile regression estimates:

```
data bp; infile '/folders/myfolders/bloodpressure.dat'
    firstobs=2; input spb spa spd dpb dpa dpd; run;
title 'Regression for Median of Systolic After';
proc quantreg data=bp; model spa=spb; run;
title 'Regression for .2 Quantile of Systolic After';
proc quantreg data=bp; model spa=spb/quantile=.2;
```

Example 10.2.2: Bootstrapping is done via a macro:

```
*Documentation at http://support.sas.com/kb/24/982.html ,code;
*support.sas.com/kb/24/addl/fusion_24982_1_jackboot.sas.txt ;
*This works similarly to the R bootstrap. You need to tell the;
*macro what statistic to use. In R this is via a function, but;
*in SAS this is via a macro.;
%include "/folders/myfolders/fusion_24982_1_jackboot.sas.txt" ;
```

and to use it, analogously with the R function boot, one writes a macro to do the calculations for each bootstrap samples. This macro has two arguments, the input data set and the output data set. These are referred to in the code within the macro by preceding their names by &:

```
%macro analyze(data=,out=);
    proc univariate data=&data noprint; var nails;
        output median=median qrange=iqr out=&out; run;
%mend;
```

where, unlike R, you need to name your macro **analyze**. Run this on the observed data set to see what you get:

```
%analyze(data=arsenic,out=results);
proc print data=results; run;
```

and then apply the macro that calls it 999 times on bootstrap samples

```
%boot(data=arseniarsenic,samples=999);
```

and after creating the bootstrap samples, evaluate the confidence interval using %bootci. This macro has at least two arguments. The first is a keyword. Use Hybrid for the residual method, pctl for the percentile method, t for the Studentized method, and BCa for BCa. Do not put quotes around the keyword. Second is the name of the variable in the data set passed from the sampler to the interval creator giving the name of the statistic on which to do calculations. Third (if present, and using argument name **student**) is the divisor for Studentization. This quantity must be a variable in the data set created by %analyze.

```
* Suppress printing with arg. print=0.;
%bootci(pctl,stat=median);
%bootci(Hybrid,stat=median);
%bootci(BCa,stat=median);
%bootci(t,stat=median,student=iqr);
```

Example 10.4.1: Again we'll use the SAS input file as above. This works similarly to the R jackknife. You need to tell the macro what statistic to use. In R this is via a function, but in SAS this is via a macro.

```
%macro analyze(data=,out=);
    proc means noprint median data=&data vardef=n;
        output out=&out(drop=_freq_ _type_) median=med ;
        var nails ;
        %bystmt;
    run;
%mend;
%jack(data=arsenic,chart=0)
```

B

Construction of Heuristic Tables and Figures Using R

Some of the preceding tables and figures were presented, not as tools in the analysis of specific data sets, but as tools for comparing and describing various statistical methods. Calculations producing these tables and figures in R are given below. Some other of the preceding tables and figures were specific to certain data sets, but are of primarily heuristic value and are not required for the data set to which they apply. Commands to produce these tables and figures are also given below. Before running these commands, load two R packages, MultNonParam from CRAN, and NonparametricHeuristic from github:

```
install.packages(c("MultNonParam","devtools"))
library(devtools)
install_github("kolassa-dev/NonparametricHeuristic")
library(MultNonParam); library(NonparametricHeuristic)
```

Figure 1.1:

```
fun.comparedensityplot()
```

Table 2.1:

```
library(VGAM)#vgam gives laplace distribution
library(BSDA)#Library gives z test
#Below the shift by -.5 for the uniform centers it at 0.
level<-fun.comparepower(sampsz=c(10,17,40),nsamp=100000,
   dist=list("rnorm","rcauchy","rlaplace","runif"),
   hypoth=c(0,0,0,-.5),alternative=c("two.sided","greater"))
print(level)
```

Figure 2.1:

```
fun.studentizedcaucyplot(10,10000)
```

Table 2.2:

```
fun.achievable()
```

Figure 2.2:

```
drawccplot()
```

Table 2.3:

```
mypower<-array(NA,c(4,3,3))
nobsv<-c(10,17,40)
for(jj in seq(length(nobsv))){
   temp<-fun.comparepower(samp1sz=nobsv[jj], nsamp=100000,
      dist=list("rnorm","rcauchy","rlaplace"),
      hypoth=(1.96+.85)*c(1,sqrt(2),1)/sqrt(nobsv[jj]))
   if(jj==1) dimnames(mypower)<-list(dimnames(temp)[[1]],
      dimnames(temp)[[2]],as.character(nobsv))
   mypower[,,jj]<-temp[,,1,1,1]
}
cat("\nPower for T and Sign Tests \n")
print(mypower)
```

Table 2.4:

```
library(VGAM); testare(.5)
```

Table 3.4:

```
fun.comparepower(samp1sz=10,samp2sz=10,altvalue=0)
```

Table 3.5:

```
fun.comparepower(samp1sz=10,samp2sz=10,altvalue=1)
```

Figures 4.1 and 4.2:

```
showmultigroupscoretest(c(3,3,4))
showmultigroupscoretest(c(3,3,4),fun.givescore(1:10,sv="ns"))
```

Figure 4.3:

```
powerplot()
```

Figure 5.1:

```
hodgeslehmannexample()
```

Figure 6.2:

```
x<-(-10):10; y<-x ; z<-5*atan(x)
plot(range(x),range(c(y,z)),type="n")
points(x,y,pch=1); points(x,z,pch=2)
legend(0,-5,pch=1:2,legend=c("y=x","y=5 atan(x)"))
```

Figure 7.1:

```
y<-x<-rnorm(100)
coin<-rbinom(100,1,.5)
y[coin==0]<--x[coin==0]
par(oma=c(0,0,3,0))
par(mfrow=c(2,2))
p1<-qqnorm(x,main="Marginal Distribution for X")
p2<-qqnorm(y,main="Marginal Distribution for Y")
p3<-qqnorm((x+y)/sqrt(2),
   main="Marginal Distribution for (X+Y)/sqrt(2)")
p4<-qqnorm((x-y)/sqrt(2),
   main="Marginal Distribution for (X-Y)/sqrt(2)")
```

Tables 7.1 and 7.2:

```
cat('\n Size of Multivariate Tests\n')
library(ICSNP)
fun.comparepower(c(20,40),ndim=2,nsamp=20000,
   dist=c("rnorm","rcauchy","rlaplace"))
cat('\n Power of Multivariate Tests\n')
fun.comparepower(c(20,40),ndim=2,nsamp=20000,
   dist=c("rnorm","rcauchy","rlaplace"),hypoth=.5)
```

Table 9.1:

```
distv<-c("rnorm","rcauchy","rlaplace","runif","rexp")
t1<-fun.testreg(dists=distv)
t2<-fun.testreg(dists=distv,npergp=50)
```

Warnings in the calculation of t1 note non-uniqueness of the quantile regression solutions.

Table 10.1:

```
fun.testboot(function(x,index) return(median(x[index])),
   alpha=.1,sampsize=20, mcsamp=1000,
   dists=c("rexp","runif","rcauchy"),true=c(log(2),.5,0))
```

Table 10.2:

```
library(bootstrap)#To give the jackknife function.
testjack(list(mean,median,mean),dists=list(rexp,rnorm),
   others=list(NULL,NULL,list(trim=0.25)),nsamp=100000)
```

Bibliography

Abramovitch, L. and K. Singh (1985, 03). Edgeworth corrected pivotal statistics and the bootstrap. *The Annals of Statistics 13*(1), 116–132.

Andrews, D. F. and A. M. Herzberg (1985). *Data: A Collection of Problems from Many Fields for the Student and Research Worker*. New York: Springer–Verlag.

Arbuthnott, J. (1712). An argument for divine providence, taken from the constant regularity observed in the births of both sexes. *Philosophical Transactions of the Royal Society of London 7*, 186–190. Reprinted in Kendall and Plackett (1977).

Benard, A. and P. V. Elteren (1953). A generalization of the method of m rankings. *Proceedings of the Koninklijke Nederlandse Akademie van Weteschappen. Series A 56 (Indagiones Mathematicae) 15*, 358–369.

Bennett, B. M. (1965, Dec). On multivariate signed rank tests. *Annals of the Institute of Statistical Mathematics 17*(1), 55–61.

Best, D. J. and P. G. Gipps (1974). Algorithm as 71: The upper tail probabilities of Kendall's tau. *Journal of the Royal Statistical Society. Series C (Applied Statistics) 23*(1), 98–100.

Best, D. J. and D. E. Roberts (1975). Algorithm as 89: The upper tail probabilities of Spearman's rho. *Journal of the Royal Statistical Society. Series C (Applied Statistics) 24*(3), 377–379.

Best, M. J. and N. Chakravarti (1990). Active set algorithms for isotonic regression: A unifying framework. *Mathematical Programming 47*, 425–439.

Bickel, P. J. (1965, 02). On some asymptotically nonparametric competitors of Hotelling's T^2. *The Annals of Mathematical Statistics 36*(1), 160–173.

Brunk, H. D. (1955, 12). Maximum likelihood estimates of monotone parameters. *The Annals of Mathematical Statistics 26*(4), 607–616.

Chen, X. and J. Kolassa (2018). Various improved approximations to distributions of quadratic test statistics for dependent rank sums. *Biomedical J. of Scientific and Technical Research 9*.

Cleveland, W. S. (1979). Robust locally weighted regression and smoothing scatterplots. *Journal of the American Statistical Association* 74(368), 829–836.

Cleveland, W. S. and S. J. Devlin (1988). Locally weighted regression: An approach to regression analysis by local fitting. *Journal of the American Statistical Association* 83(403), 596–610.

Clopper, C. J. and E. S. Pearson (1934, 12). The use of confidence or fiducial limits illustrated in the case of the binomial. *Biometrika* 26(4), 404–413.

Conover, W. J. and R. L. Iman (1979). On multiple-comparisons procedures. Technical report, Los Alamos Scientific Laboratory, Los Alamos, NM.

Cox, D. R. and E. J. Snell (1981). *Applied Statistics: Principles and Examples*. New York: Chapman and Hall.

Cramér, H. (1946). *Mathematical Methods of Statistics*. Princeton University Press.

David, S. T., M. G. Kendall, and A. Stuart (1951). Some questions of distribution in the theory of rank correlation. *Biometrika* 38(1/2), 131–140.

Davison, A. C. and D. V. Hinkley (1997). *Bootstrap Methods And Their Application* (first ed.). Cambridge Series in Statistical and Probabilistic Mathematics. Cambridge University Press.

Dinneen, L. C. and B. C. Blakesley (1973). Algorithm as 62: A generator for the sampling distribution of the Mann-Whitney U statistic. *Journal of the Royal Statistical Society. Series C (Applied Statistics)* 22(2), 269–273.

Dunn, O. J. (1964). Multiple comparisons using rank sums. *Technometrics* 6(3), 241–252.

Dwass, M. (1956, 06). The large-sample power of rank order tests in the two-sample problem. *The Annals of Mathematical Statistics* 27(2), 352–374.

Dwass, M. (1985). On the convolution of Cauchy distributions. *The American Mathematical Monthly* 92(1), 55–57.

Dykstra, R. L. (1981). An isotonic regression algorithm. *Journal of Statistical Planning and Inference* 5, 355–363.

Edgeworth, F. Y. (1893). Viii. exercises in the calculation of errors. *The London, Edinburgh, and Dublin Philosophical Magazine and Journal of Science* 36(218), 98–111.

Efron, B. (1981). Nonparametric standard errors and confidence intervals. *Canadian Journal of Statistics* 9(2), 139,158.

Efron, B. (1987). Better bootstrap confidence intervals. *Journal of the American Statistical Association 82*(397), 171–185.

Efron, B. and R. J. Tibshirani (1993). *An Introduction to the Bootstrap.* New York: Chapman and Hall.

El Maache, H. and Y. Lepage (2003). Spearman's rho and Kendall's tau for multivariate data sets. In M. Moore, S. Froda, and C. Léger (Eds.), *Mathematical Statistics and Applications: Festschrift for Constance van Eeden*, Volume 42 of *Lecture Notes–Monograph Series*, Beachwood, OH, pp. 113–130. Institute of Mathematical Statistics.

Epanechnikov, V. (1969). Non-parametric estimation of a multivariate probability density. *Theory of Probability & Its Applications 14*(1), 153–158.

Erdös, P. and A. Réyni (1959). On the central limit theorem for samples from a finite population. *Publications of the Mathematical Institute of the Hungarian Academy of Sciences 4*, 49–61.

Fan, J. (1992). Design-adaptive nonparametric regression. *Journal of the American Statistical Association 87*(420), 998–1004.

Festinger, L. (1946). The significance of difference between means without reference to the frequency distribution function. *Psychometrika 11*(2), 97–105.

Fieller, E. C. (1954). Some problems in interval estimation. *Journal of the Royal Statistical Society. Series B (Methodological) 16*(2), 175–185.

Fisher, R. A. (1925). *Statistical Methods for Research Workers.* Edinburgh: Oliver and Boyd.

Fisher, R. A. (1926). On a distribution yielding the error functions of several well known statistics. In *Proceedings of the International Mathematical Congress*, pp. 805–813. Congress ran 1924. Fisher, Statistical Methods for Research Workers, dates it to 1924. Hald dates to 1928. Addendum in collected papers is dated 1927.

Fisher, R. A. (1930). *Statistical Methods for Research Workers* (Third ed.). Edinburgh: Oliver and Boyd.

Fisher, R. A. (1973). *Statistical Methods for Research Workers* (Fourteenth (reprinted) ed.). Edinburgh: Oliver and Boyd.

Friedman, M. (1937). The use of ranks to avoid the assumption of normality implicit in the analysis of variance. *Journal of the American Statistical Association 32*(200), 675–701.

Hájek, J. (1960). Limiting distributions in simple random sampling from a finite population. *Publications of the Mathematical Institute of the Hungarian Academy of Sciences 5*, 361–374.

Hald, A. (1998). *A History of Mathematical Statistics from 1750 to 1930.* John Wiley and Sons, Inc.

Hall, P. (1988, 10). Rate of convergence in bootstrap approximations. *The Annals of Probability 16*(4), 1665–1684.

Hall, P. (1992). *The Bootstrap and Edgeworth Expansion.* Springer-Verlag.

Haynam, G. E., Z. Govindarajulu, F. C. Leone, and P. Siefert (1982a). Tables of the cumulative non-central chi-square distribution - part 1. *Series Statistics 13*(3), 413–443.

Haynam, G. E., Z. Govindarajalu, F. C. Leone, and P. Siefert (1982b). Tables of the cumulative non-central chi-square distribution - part 2. *Series Statistics 13*(4), 577–634.

Hettmansperger, T. and J. McKean (2011). *Robust Nonparametric Statistical Methods.* Boca Raton: CRC Press.

Hettmansperger, T. P. (1984). *Statistical Inference Based on Ranks.* Melbourne, FL: Krieger.

Higgins, J. J. (2004). *Introduction to Modern Nonparametric Statistics.* Cengage Learning.

Hoaglin, D. C. and R. E. Welsch (1978). The hat matrix in regression and anova. *The American Statistician 32*(1), 17–22.

Hodges, J. L. and E. L. Lehmann (1963). Estimates of location based on rank tests. *The Annals of Mathematical Statistics 34*(2), 598–611.

Hoeffding, W. (1948, 09). A class of statistics with asymptotically normal distribution. *The Annals of Mathematical Statistics 19*(3), 293–325.

Hotelling, H. (1931, 08). The generalization of student's ratio. *The Annals of Mathematical Statistics 2*(3), 360–378.

Hotelling, H. and M. R. Pabst (1936, 03). Rank correlation and tests of significance involving no assumption of normality. *The Annals of Mathematical Statistics 7*(1), 29–43.

Huber, P. J. (1964, 03). Robust estimation of a location parameter. *The Annals of Mathematical Statistics 35*(1), 73–101.

Johnson, N. L., S. Kotz, and N. Balakrishnan (1995). *Continuous Univariate Distributions* (Second ed.), Volume 2. Wiley-Interscience.

Jonckheere, A. R. (1954). A distribution-free k-sample test against ordered alternatives. *Biometrika 41*(1/2), 133–145.

Bibliography

Kaarsemarker, L. and A. van Wijngaarden (1953). Tables for use in rank correlation. *Statistica Neerlandica* 7, 41–54.

Kawaguchi, A., G. G. Koch, and X. Wang (2011). Stratified multivariate Mann–Whitney estimators for the comparison of two treatments with randomization based covariance adjustment. *Statistics in Biopharmaceutical Research* 3(2), 217–231.

Kendall, M. G. (1938). A new measure of rank correlation. *Biometrika* 30(1-2), 81–93.

Koenker, R. (2000). Galton, Edgeworth, Frisch, and prospects for quantile regression in econometrics. *Journal of Econometrics* 95(2), 347 – 374.

Köhler, M., A. Schindler, and S. Sperlich (2014). A review and comparison of bandwidth selection methods for kernel regression. *International Statistical Review* 82(2), 243–274.

Kolassa, J. E. (2006). *Series Approximation Methods in Statistics, 3rd Edn.* New York: Springer – Verlag.

Kolassa, J. E. and Y. Seifu (2013). Nonparametric multivariate inference on shift parameters. *Academic Radiology* 20(7), 883 – 888.

Kramer, C. Y. (1956). Extension of multiple range tests to group means with unequal numbers of replications. *Biometrics* 12(3), 307–310.

Kramer, C. Y. (1957). Extension of multiple range tests to group correlated adjusted means. *Biometrics* 13(1), 13–18.

Kruskal, W. H. and W. A. Wallis (1952). Use of ranks in one-criterion variance analysis. *Journal of the American Statistical Association* 47(260), 583–621.

Laplace, P. S. D. (1818). *Deuxiéme Supplńent a la Théorie Analytique des Probablilitiés*. Paris: Courcier.

Lehmann, E. L. (1953). The power of rank tests. *The Annals of Mathematical Statistics* 24(1), 23–43.

Lehmann, E. L. (1993). The Fisher, Neyman-Pearson theories of testing hypotheses: One theory or two? *Journal of the American Statistical Association* 88(424), 1242–1249.

Lehmann, E. L. (2006). *Nonparametrics: Statistical Methods Based on Ranks* (First edition revised ed.). Springer.

Mann, H. B. and D. R. Whitney (1947, 03). On a test of whether one of two random variables is stochastically larger than the other. *The Annals of Mathematical Statistics* 18(1), 50–60.

Miller, R. G. (1974). The jackknife–a review. *Biometrika* 61(1), 1–15.

Mood, A. M. (1950). *Introduction to the Theory of Statistics* (First ed.). New York: McGraw Hill.

Nadaraya, E. A. (1964). On estimating regression. *Theory of Probability & Its Applications 9*(1), 141–142.

Nemenyi, P. (1963). Distribution-Free Multiple Comparisons. Ph. D. thesis, Princeton University.

Neyman, J. and E. S. Pearson (1933). On the problem of the most efficient tests of statistical hypotheses. *Philosophical Transactions of the Royal Society of London. Series A, Containing Papers of a Mathematical or Physical Character 231*, 289–337.

Noether, G. E. (1950, 06). Asymptotic properties of the Wald-Wolfowitz test of randomness. *The Annals of Mathematical Statistics 21*(2), 231–246.

Oja, H. and R. H. Randles (2004, 11). Multivariate nonparametric tests. *Statist. Sci. 19*(4), 598–605.

Page, E. B. (1963). Ordered hypotheses for multiple treatments: A significance test for linear ranks. *Journal of the American Statistical Association 58*(301), 216–230.

Pearson, E. S. (1931). The analysis of variance in cases of non-normal variation. *Biometrika 23*(1/2), 114–133.

Pearson, K. (1907). *On Further Methods of Determining Correlation*. Drapers' company research memoirs. Biometric ser. IV. London: Dulau and Co.

Pitman, E. J. G. (1948). Lectures on nonparametric statistical inference. Columbia University.

Prentice, M. J. (1979). On the problem of m incomplete rankings. *Biometrika 66*(1), 167–170.

Quenouille, M. H. (1956). Notes on bias in estimation. *Biometrika 43*(3/4), 353–360.

R Core Team (2018). *R: A Language and Environment for Statistical Computing*. Vienna, Austria: R Foundation for Statistical Computing.

Randles, R. H. (1989). A distribution-free multivariate sign test based on interdirections. *Journal of the American Statistical Association 84*(408), 1045–1050.

Rosenblatt, M. (1956, 09). Remarks on some nonparametric estimates of a density function. *The Annals of Mathematical Statistics 27*(3), 832–837.

Sankaran, M. (1963, 06). Approximations to the non-central chi-square distribution. *Biometrika 50*(1-2), 199–204.

Bibliography

SAS Institute Inc. (2017). *SAS/STAT 14.3 User's Guide*. Cary, NC: SAS Institute Inc.

Savage, I. R. (1956, 09). Contributions to the theory of rank order statistics-the two-sample case. *The Annals of Mathematical Statistics 27*(3), 590–615.

Savitzky, A. and M. J. E. Golay (1964). Smoothing and differentiation of data by simplified least squares procedures. *Analytical Chemistry 36*(8), 1627–1639.

Schoenberg, I. J. (1946). Contributions to the problem of approximation of equidistant data by analytic functions. Part A. On the problem of smoothing or graduation. A first class of analytic approximation formulae. *Quarterly of Applied Mathematics 4*, 46–99.

Scott, D. W. (1979, 12). On optimal and data-based histograms. *Biometrika 66*(3), 605–610.

Sen, P. K. (1968). Estimates of the regression coefficient based on Kendall's tau. *Journal of the American Statistical Association 63*(324), 1379–1389.

Shao, J. and D. Tu (1995). *The Jackknife and Bootstrap* (first ed.). Springer Series in Statistics. New York: Springer-Verlag.

Sheather, S. J. and M. C. Jones (1991). A reliable data-based bandwidth selection method for kernel density estimation. *Journal of the Royal Statistical Society. Series B (Methodological) 53*(3), 683–690.

Silverman, B. W. (1986). *Density Estimation for Statistics and Data Analysis*. Monographs on Statistics and Applied Probability. London: Chapman and Hall.

Skillings, J. H. and G. A. Mack (1981). On the use of a Friedman-type statistic in balanced and unbalanced block designs. *Technometrics 23*(2), 171–177.

Small, C. G. (1990). A survey of multidimensional medians. *International Statistical Review / Revue Internationale de Statistique 58*(3), 263–277.

Snedecor, G. W. (1934). *Calculation and Interpretation of Analysis of Variance and Covariance*. Collegiate Press, Inc.

Spearman, C. (1904). The proof and measurement of association between two things. *The American Journal of Psychology 15*(1), 72–101.

Stieltjes, T. J. (1894). Recherches sur les fractions continues. *Annales de la Faculté des Sciences de Toulouse 8*, J1–J122.

Stigler, S. M. (1973). Studies in the history of probability and statistics. xxxii: Laplace, Fisher and the discovery of the concept of sufficiency. *Biometrika 60*(3), 439–445.

Stigler, S. M. (1984, 12). Studies in the history of probability and statistics xl boscovich, simpson and a 1760 manuscript note on fitting a linear relation. *Biometrika* 71(3), 615–620.

Stigler, S. M. (1986). *The History of Statistics: The Measurement of Uncertainty Before 1900*. Cambridge: Harvard University Press.

Student (1908). Probable error of a correlation coefficient. *Biometrika* 6(2/3), 302–310.

Sturges, H. A. (1926). The choice of a class interval. *Journal of the American Statistical Association* 21(153), 65–66.

Terpstra, T. J. (1952). The asymptotic normality and consistency of Kendall's test against trend, when ties are present in one ranking. *Indagationes Mathematicae* 14, 327–333.

Theil, H. (1950). A rank-invariant method of linear and polynomial regression analysis. i. *Indagationes Mathematicae* 12, 85–91.

Tramo, M., W. Loftus, T. Stukel, R. Green, J. Weaver, and M. Gazzaniga (1998, May). Brain size, head size, and intelligence quotient in monozygotic twins. *Neurology* 50(5), 1246-52.

Tukey, J. W. (1949). Comparing individual means in the analysis of variance. *Biometrics* 5(2), 99–114.

Tukey, J. W. (1953). The problem of multiple comparisons : introduction and parts a, b, and c. Photocopy of typescript from Princeton University eventually published in 1993.

Tukey, J. W. (1993). Reminder sheets for "allowances for various types of error rates". In H. I. Braun (Ed.), *The Collected Works of John W. Tukey Volume VIII: Multiple Comparisons, 1948-1983*, pp. 335–339. New York, NY: Chapman and Hall.

van de Wiel, M. and A. Di Bucchianico (2001). Fast computation of the exact null distribution of Spearman's rho and Page's L statistic for samples with and without ties. *Journal Of Statistical Planning And Inference* 92(1-2), 133–145.

Waerden, B. L. V. D. (1952). Order tests for the two-sample problem and their power. *Indagationes Mathematicae* 14, 453–458.

Watson, G. S. (1964). Smooth regression analysis. *Sankhyā: The Indian Journal of Statistics, Series A* 26(4), 359–372.

Westenberg, J. (1948). Significance test for median and interquartile range in samples from continuous populations of any form. In *Proceedings for Koninklijke Nederlandse Akademie van Wetenschappen*, pp. 252–261. Koninklijke Nederlandse Akademie van Wetenschappen.

Wilcoxon, F. (1945). Individual comparisons by ranking methods. *Biometrics Bulletin 1*(6), 80–83.

Yarnold, J. K. (1972). Asymptotic approximations for the probability that a sum of lattice random vectors lies in a convex set. *The Annals of Mathematical Statistics 43*(5), 1566–1580.

Zhong, D. and J. Kolassa (2017, September). Moments and Cumulants of The Two-Stage Mann-Whitney Statistic. *ArXiv e-prints*.

Index

alternative hypothesis, 7
analysis of variance, 69–72, 74, 75, 77, 79, 81, 102, 180
Ansari-Bradley test, 58, 63
asymptotic relative efficiency, 29

bandwidth, 145
basic bootstrap confidence interval, 171
bias corrected and accelerated, 172
binomial distribution, 5, 20, 24, 144, 146
binomial test, 20
Bonferroni procedure, 71, 82, 107
bootstrap, 167–181
bootstrap sample, 168

Cauchy distribution, 3, 16, 22, 23, 32, 53, 135, 159, 169, 175
central limit theorem, 15, 19, 42, 48, 49, 70, 115, 134, 150, 167
chi-square distribution, 5, 70, 72–77, 86, 87, 92, 103–105, 130, 134, 136
concordant, 118, 120
contrast, 71, 83, 91
correction for continuity, 22, 29, 45, 49, 50, 55, 77
Cramér-von Mises test, 63, 64
critical region, 7, 9, 10, 77
critical value, 7–10, 20, 21, 24, 28–30, 49, 62, 76, 77, 87, 90, 91, 100, 117

degrees of freedom, 5
discordant, 118, 120
double exponential distribution, 3

Edgeworth approximation, 116, 122, 169
efficacy, 31–33, 56–58, 71, 91
Epanechnikov kernel, 145
exponential distribution, 4, 159, 175, 184
exponential scores, 50

F distribution, 5, 6, 70, 130
family-wise error rate, 71, 72

Gaussian distribution, 1
generalized sign test, 27

H test, see Kruskal-Wallis test
Hermite polynomials, 116
histogram, 143, 144, 176
Hodges-Lehmann estimator, 61, 99
Honest Significant Difference, 72, 82, 107

isotonic regression, 155

jackknife, 181–184
Jonckheere-Terpstra test, 82–85, 87, 91

Kendall's τ, 119–122, 124, 125
kernel, 144, 145, 147, 150, 151
kernel density estimation, 144, 145, 147, 150
kernel smoothing, 150, 151, 153, 154
Kolmogorov-Smirnov test, 63, 64
Kruskal-Wallis test, 76–78, 81, 82, 84, 86–89, 92, 103, 104, 133

L^1 regression, 158–162
Laplace distribution, 3, 16, 22, 23, 32, 57, 58, 135, 159

211

Least Significant Difference, 71
Lehmann alternative, 50, 58
locally weighted regression, 153, 154
location-scale family, 4
logistic distribution, 4, 50, 56, 57, 85, 89

Mann-Whitney test, 48–50, 55–58, 60–62, 81, 83, 86, 102, 103, 133
Mood's median test, 43–46, 50, 53, 54, 60, 141
multinomial distribution, 5, 74
multivariate Gaussian distribution, 113, 115, 121, 129
multivariate median, 130, 132

Nadaraya-Watson smoothing, 150, 153, 154
non-central Cauchy distribution, 3
non-central chi-square distribution, 6, 86, 89, 92
non-centrality parameter, 6
normal distribution, 1
normal scores, 50, 53, 79
null hypothesis, 6

one-sided hypothesis, 7
order statistics, 26, 50–52, 57, 62

p-value, 10
Page's test, 108, 109
paired t-test, 95
parametric bootstrap, 169
Pearson correlation, 113–116, 120, 121, 124, 176
percentile method, 171
permutation test, 52
pivot, 12, 177, 178
pivotal, 177
pooled adjacent violators, 155
power, 8, 10, 11, 22, 23, 28–33, 43, 53–55, 58, 69, 71, 84–87, 89–92, 102, 103, 135, 139
Prentice test, 106

quantile regression, 158–162

random X bootstrap, 175
relative efficiency, 28, 29, 31, 32, 55–57, 84, 91, 101
resampling distribution, 168
resistant, 163

Savage scores, 50, 51, 53, 78, 79, 107
Siegel-Tukey test, 58, 59
sign test, 20, 22–25, 27, 31–33, 43, 62, 96, 97, 105, 132–136
Spearman correlation, 116, 118–121, 123, 124
Spearman Rank correlation, 115
spline, 155, 157
standard Gaussian distribution, 2
standard normal distribution, 2
Student's t distribution, 6, 16
Studentized, 177
Studentized range distribution, 72, 81

test level, 7, 9–11, 16, 17, 20–23, 28, 31, 116
test statistic, 7
tied observations, 54, 107
two-sample pooled t statistic, 40
two-sample pooled t-test, 40, 41, 51, 52, 71, 72, 81
two-sided hypothesis, 9
type I error rate, 7
type II error rate, 8

U statistic, 48
uniform distribution, 2, 3, 16, 150
usual method, 171

van der Waerden scores, 50, 51, 78

Walsh averages, 99–101
Wilcoxon rank-sum test, 46–50, 53–58, 60–62, 81, 83, 86, 99, 138
Wilcoxon signed-rank test, 96, 99–101, 132–136